WIRING A HOUSE

WIRING A HOUSE

Completely Revised and Updated

REX CAULDWELL

The Taunton Press

Text and photographs © 2002 by Rex Cauldwell

Illustrations © 2002 by The Taunton Press, Inc.

 The Taunton Press
Inspiration for hands-on living®

The Taunton Press, Inc., 63 South Main Street, PO Box 5506, Newtown, CT 06470-5506
e-mail: tp@taunton.com
Distributed by Publishers Group West

Jacket/cover design: Cathy Cassidy
Interior design and layout: Jeff Potter/Potter Publishing Studio
Illustrator: Gayle Rolfe
Photographer: Rex Cauldwell, unless otherwise noted

For Pros/by Pros® is a trademark of The Taunton Press, Inc.,
registered in the U.S. Patent and Trademark Office.

The following manufacturers/names appearing in *Wiring a House* are trademarks: 3M™, Above Code™,
Adjust-A-Box ™, Black & Decker®, Bosch®, Bussman®, Burndy®, Carlon®, DeWalt®, EZ Wire®,
Fluke®, Gardner Bender®, GE®, Gen/Tran®, Greenlee® , Hitachi®, Honda®, Hunter®, Ideal®,
Lennox®, Lutron®, Makita®, Milwaukee®, Nail Eater II®, Primax™, QO®, Rangemaster®, Romex®,
Sawzall®, Scotch®, Siemens®, Simpson 260®, Slug-Buster®, SureTest®, Torx®, Wire-Nut®.

Library of Congress Cataloging-in-Publication Data

Cauldwell, Rex.
    Wiring a house / Rex Cauldwell.— Completely rev. and updated
        p. cm. — (For pros, by pros)
Includes index.
    ISBN 1-56158-527-0
  1. Electric Wiring, Interior. I. Title. II. Series.
    TK3285 .C38 2002
    621.319′24—dc21

2002001542

Printed in the United States of America

10 9 8 7 6 5 4 3

About Your Safety: Wiring is inherently dangerous. We try to promote safe work practices
throughout this book, but what is safe for one builder or homeowner under certain circumstances
may not be safe for you under different circumstances. So don't try anything you learn about here
(or elsewhere) unless you're certain that it's safe for you. If something doesn't feel right, don't do it.
Look for another way. Please keep safety foremost in your mind whenever you're working.

To my wife, Diana, who kept the household together while I wrote this book. Without the skilled help of my grand-daughters, Katy and Elizabeth, who were always demanding to use the computer, and the specialized assistance of my cats, Pie Face and Birdy, who were always getting themselves tied up in the computer cords, I could have finished this book at least four times faster.

ACKNOWLEDGMENTS

No one person creates a book. It is the accumulation of many hands and minds. At The Taunton Press I would like to thank Julie Trelstad, who guided me through the conception of the original edition, and Tom McKenna, who led me through the editing process. I cannot imagine two better people to work with—they made this book enjoyable. Thanks also to Joe Hurst-Wajszczuk and Peter Chapman for their work on this revision.

A work of this kind is very technical, and it is very easy for errors to slip by. In addition, the way houses are wired on the East Coast may not be the same as on the West Coast. I would like to thank the technical readers, who helped minimize these problems: Bill Goode, the toughest electrical inspector that Roanoke County, Virginia, ever had; April Elkin, of Local #637 in Roanoke, Virginia; and Redwood Kardon, building inspector for the city of Oakland, California.

Many companies helped with this book by providing photographic material and technical information about their products. I would like to give heartfelt thanks for their help to Vivian Beaulieu of Marchant and Field; Ben Bird of Certified Insulated Products; Doug Kirk of King Technology, Siemens; Constance Malenfont of TayMac Corp.; Jean Miskimon of DeWalt public relations; M. F. Reed of Tytewadd Power Filters; and Jay Thomas of AFC Cable Systems. I would also like to thank the good people at Carlon®; Bussman; Greenlee Textron, Inc.; Hunter Fan Company; Hubbell Incorporated; Ideal Industries, Inc.; Lutron; Magnus Industries, Inc.; and Technology Research Corp.

Contents

Preface to the Revised Edition

There are two constants in life—paying taxes and changing codes. The purpose of this preface is to assure my many fans and avid readers that this new edition takes into consideration all the codes that have changed since I originally wrote *Wiring a House*. Here are a few of the more common code changes:

- One expected change was in marking the white insulated wire in switch legs. At the time *Wiring a House* was written, code did not require the installer to indicate whether a white insulated wire, such as a traveler in a 3- or 4-way switching circuit, was not a neutral or whether a common switch was using a white wire for its incoming power. Code now requires the white insulated wire to be taped (any tape color except white, gray, or green) to make sure you know that it is not a neutral.

- One change that I particularly dislike is the new rule on bathroom receptacles. Now you can put a whirlpool tub, a bath heater, a heated towel rack, bath lights, or whatever on a single bathroom receptacle circuit as long as that circuit stays in that one bath.

- Outbuildings, if they are wired, are required to use ground-fault circuit interrupters (GFCIs) whether they are finished or unfinished.

- To keep children from grabbing appliance cords and pulling the appliance down onto themselves, all countertop receptacles are now required to be above the countertop. Exceptions are made for the handicapped.

- Installers are now allowed to use nonmetallic (NM) cable, in residential installations, at any grade level. Previously, code had allowed NM for up to three floors, but there was some debate about whether or not you could count the basement as one story and the attic as another.

- AFCIs (arc-fault circuit interrupters) are now required for any circuit in the bedroom. That includes receptacles, lighting circuits, smoke alarms, and so on.

- Dedicated receptacles in unfinished basements that feed fire and burglar alarms are not required to have GFCI protection.

- All 15- and 20-amp 125-volt receptacles installed outside (you are required to have one in the front and one in the back) must utilize waterproof covers that keep the receptacle waterproof even when a cord is inserted.

ABOVE CODE One major change in this book's revision is the addition of my Above Code™ wiring system throughout the book. Above Code is my way of ensuring that the homeowner has a safe, high-quality, trouble-free, long-lived electrical installation. An example of using Above Code is in wiring bathroom receptacles. If you were to simply "meet code," you could load up a bath circuit with the lights, fan, heater, and so on. A single hair dryer on a low setting would wind up kicking the breaker and leaving you in the dark. My Above Code system shows you how a bath *should* be wired so that doesn't happen. In addition, my system tells you how to wire better, what products to buy, and which ones to avoid.

This revision has a number of other changes. I've replaced black-and-white photos with color photos to give you a better understanding of the material. I've reworked artwork—and even complete chapters—that didn't have practical applications in favor of adding information that you could really use. I've also added color to the line art to give a greater sense of realism. To aid clarity, I added the chapter Wiring Room by Room (p. 115) so you wouldn't have to hunt all over the book to find the specs to wire a specific room. Due to their recent popularity, I've also added chapters on home standby generators and lightning protection.

I think you will be pleased with my new edition of *Wiring a House*. New buyers, I hope, will find this edition to be their "real-life" wiring reference—the one with the dog-eared pages—while all those written by desk jockeys will gather dust on the shelf. For those who already own a copy of the earlier edition, I hope you will find this edition even more useful and will add it to your library as well.

—*Rex Cauldwell*

Introduction

My family has three generations of electricians—I am the third. As a kid, I remember working in spooky old buildings that had been around since the Civil War. The attics and basements of those houses were especially scary to me, but it was the crawl spaces I hated most. While lying in those damp or dust-choked caves, I remember pulling wires through floor joists spanned by spider webs clogged with insect carcasses, attempting in vain to ignore the multilegged thing crawling up my leg—all the while trying not to knock my head against another darn joist and hoping the flashlight wouldn't die. Those are not fond memories, but I learned a lot. And it was literally from the ground up that I was taught about electricity and wiring a house. Now I'm both a master electrician and a master plumber and have my own company.

I have written this book from lifelong experience and knowledge, some of which has been passed down through each generation. However, there is no one within my family to pass the gauntlet to—no fourth generation to pick up the trade. Therefore, by reading this book, you will become heir to my knowledge and experience. You, in effect, will become the fourth generation.

Three main themes of the book are safety, design, and materials. If you're doing the work yourself, I'll tell you how to do it safely. If you're not doing the work, you'll gain enough information from these pages to ask educated questions, to understand what an electrician is doing, and to know whether he or she is doing it correctly. You'll also learn what makes a good electrical design and how to choose the best materials—not the lowest-cost materials—for your situation.

Safety Is Paramount

As an electrician, safety is a primary concern—both for my clients and for me. Electricity can kill, so it's very important to be alert while working with it and to use safety equipment. I've received shocks before, and I would not be here today had it not been for a GFCI-protected receptacle. *If you are not competent working around electricity, hire an electrician do the work.*

One of the aims of this book is to help you understand the basic principles of electricity so that you can give it the respect it deserves. You'll learn not only how electricity flows but also how to work according to a plan, so that any wiring job can be done safely and without fear. Throughout the book I offer safety tips that could save your life. Wiring can be tricky, so take your time and don't cut corners.

2

Tools

Tools are important to any person working in the building trades. Using the right tool for the right job will make the work go smoothly and safely. The same holds true for electrical tools; however, using the wrong tool or a low-quality tool will not only cause headaches by making the job more difficult, but it also could seriously hurt or kill you. That's why I devoted an entire chapter to tool use. In it I give you good background knowledge of electrical tools so that you will know which tools to buy, and even more important, which ones not to buy. I also illustrate how to use tools correctly, not just in the tools chapter but throughout the book.

Meeting Code Is Not Good Enough

Minimum code means exactly that—it's the absolute minimum required to pass inspection. And I'm sorry to admit that on many jobs even minimum codes sometimes aren't enforced. Most inspectors are already backlogged and overworked—they have time only to check for obvious violations. They cannot trace every wire to make sure it goes to the right location or even verify that the wire is the proper gauge. Therefore, you cannot assume that an electrical system has been installed correctly or even safely just because it has passed inspection. Sadly, the bottom line is that it's normally up to the installer or homeowner to know what needs to be done and to see that it is done correctly. Knowing this, I try to give you enough information to know right from wrong and what works and what doesn't, so that you can make intelligent decisions about the design of an electrical system.

In addition, this book will teach you how to develop a good, safe, high-quality electrical design, not one that simply meets minimum code. For example, one time I was called out to rewire a recently built house. It had passed all electrical inspections, the walls were up and painted, and the owner had moved in. The contractor had only been obligated to build to minimum specs. The outlets were spaced 12 ft. apart, with no receptacle outlet where it was needed for a specific piece of furniture. A cheap, poorly designed electrical panel that was 99%

full upon completion of the house was taking all the load it could handle, so nothing could be added (such as a spa). The electrical system—in fact, the entire house—was built without any consideration for the owner's needs. The owner had to pay twice: once to meet minimum code and a second time to get things custom-designed. A good design surpasses minimum code and takes the owner's needs into consideration. (See my Above Code reference in the Preface.)

Low-Bid Jobs Are Cheap—for a Reason

It may come as a surprise to some people that it is impossible to obtain high-quality material on a low-bid job. Why? From the contractor's viewpoint, the object of the bid is to get the job. If I were to put together a bid that includes good-quality, high-end material and my competitor makes a bid that includes cheap material, my bid will be significantly higher—and

I most likely will not get the job. For contractors, this book will illustrate when high-end material is appropriate and when you can get by with average-quality material. With this knowledge, you'll be able to put together a bid that's reasonably priced, without compromising overall quality.

As a homeowner, you should know that when you choose the lowest bid, you may get exactly what you pay for. But if you specify in advance the type and exact grade of materials you want, so that all contractors are bidding with the same standards in mind, you can choose the lowest price knowing that you haven't compromised quality. This book will give you the knowledge to make informed decisions about the wiring system in your house.

A Book Written from Experience

This book is unlike any other wiring book on the market. Written by an electrician for homeowners, do-it-yourselfers, and professionals alike, it is full of stories and experiences of exactly what happens when wiring a house. I even talk about some of the common mistakes that both pros (including myself) and do-it-yourselfers make so that you can avoid them from the outset.

I've always hated the standard how-to wiring books because, for the most part, they're not written by practicing electricians. Instead, some desk jockey rehashes stuff from other books written by other desk jockeys. These books always pick perfect textbook situations with photos taken in a studio. They never tell you the problems you will encounter and what to do when things go wrong—let alone the experiences of the authors. They can't, because the people writing the book have rarely done what they're telling you to do. I think both professionals and novices will appreciate my book because it's honest— I've done the work.

CHAPTER 1

The Basics

NOT KNOWING the basics of electricity can get you into a lot of trouble—painful trouble. I remember when my Uncle Bud started teaching me about electricity back in the '50s. I accidentally tripped on some black wires in the attic and got shocked as I grabbed onto a funny-looking porcelain knob to stop my fall. This was my first lesson in basic wiring: Don't touch this and don't trip over that.

Today, the basics are a little more complex than those by which my Uncle Bud and my Grandfather Bunt worked. (My Grandfather Bunt didn't even have a meter to test for voltage. He used an old-time method that I can't print here for fear that somebody might try it and get electrocuted.) Although they didn't have the modern test equipment that we do, they talked the same electrical language, using terms such as current flow, voltage, power, and resistance. But I'm sure they didn't understand their meanings as well as we do today. In this chapter, I'll help you make sense of the basics of electricity so you understand what you'll be doing and why.

All electrical current originates from and comes back to the three side terminals of the utility transformer.

How Electricity Flows

Wire works much like a garden hose, but instead of conveying water, it conveys electricity from one location to another. When you turn on a hose faucet, water entering from the spigot pushes on water already in the hose, which pushes water out the other end. Electricity flows in much the same way. An electron flows in one end of the wire, which knocks an electron, which in turn knocks another electron, until an electron eventually comes out the other end.

The water analogy can be used to describe the other elements of electricity. To get water to flow, we need water pressure. To get electricity to flow, we need electrical pressure. Electrical pressure, or voltage, can be provided from either an electrical utility or a battery. And just as greater water pressure means more water flow, higher voltages provide greater electrical flow. This flow is called current. With both water and electricity, the diameter of the hose or wire limits what you get out of it in a given amount of time. This flow restriction is referred to as resistance.

Elements of electricity

Electricity is the organized flow of electrons along a conductor. It is generated through heat, pressure, friction, light, chemical action, or magnetism. The four basic elements of electricity are voltage, current, resistance, and power. Ohm's Law, a basic principle of electricity, states that voltage is directly proportional to the current and resistance. You can calculate any one value if the other two are known (E=IR, where E is the voltage, I is the current, and R is the resistance).

VOLTAGE is the electrical pressure provided by a battery or other power source. The higher the voltage, the more current the source can produce. In a basic formula without numbers, voltage is abbreviated by the letter E, which stands for EMF, electromotive force. Voltage is measured in volts; a voltage quantity is always followed by a capital V or the word "volts" (120V or 120 volts). A multimeter measures the voltage between two points on a circuit, or if the current and resistance are known, it can be calculated using Ohm's Law.

Is It Wire or Cable?

The terms "wire" and "cable" are used interchangeably by most electricians. Technically, a wire is an individual conductor and a cable is composed of two or more wires surrounded by a jacket. However, most electricians still refer to cable simply as wire. When an electrician requests 12-gauge wire for a particular spot, he normally wants 12-gauge cable, unless he is running individual wires within a conduit system.

On the other hand, large-diameter wires are often referred to as cables. And the word wire can be both singular and plural. Because this gets very confusing, for clarity, it's better to say "conductor" as opposed to "wire." A conductor is always a single wire.

CURRENT is the organized flow of electrons from one point to another on a circuit. Electrons orbit around atoms just as planets orbit the sun. When an electron is knocked out of orbit and starts traveling down the circuit, it knocks other electrons out of their orbits, creating current. There are two types of current: alternating and direct (see DC and AC on p. 10).

Current can be increased by raising the voltage or by lowering the resistance. In a basic formula without any numbers, current is represented by the letter I. Current is measured in amperes, or amps (e.g., 20 amps). Electrical current can be measured by inserting an ammeter in series with the wiring (cutting the wire and attaching a meter lead to each wire end), but it is much safer to use a clamp-on meter. This type of meter clamps around the wire and measures the electromagnetic waves emitting from the wire to determine the current. If the voltage and resistance are known, current can be calculated using Ohm's Law: I=E/R.

RESISTANCE is the inherent physical opposition to current flow (like a dam trying to hold a river back). The opposition is created by electrons refusing to be stripped of their atoms and sent down

Inside the Utility-Pole Transformer

A utility-pole transformer has an insulated terminal on top, which connects a high-voltage line to a primary coil (see the drawing at right).

The primary coil is connected to ground/neutral via a center tap, which then transfers its energy into a secondary coil to step down the voltage from 7,500 volts to 240 volts.

Grounding the middle of the secondary coil splits 240 volts in half, resulting in two 120-volt sources and one 240-volt source.

CT = Center tap

H = Hot

To high-voltage line

Primary coil
7,500 volts

H
120 volts
CT
120 volts
H

Secondary coil
240 volts

240-volt Load

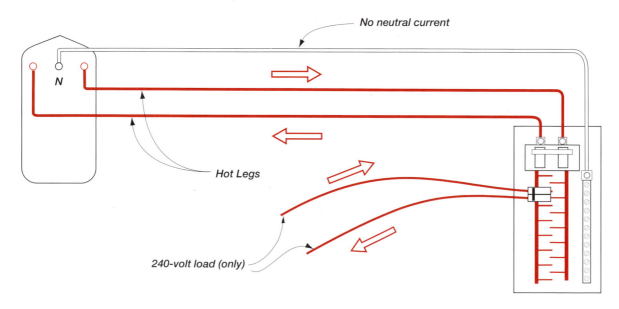

No neutral current

N

Hot Legs

240-volt load (only)

120-volt Load

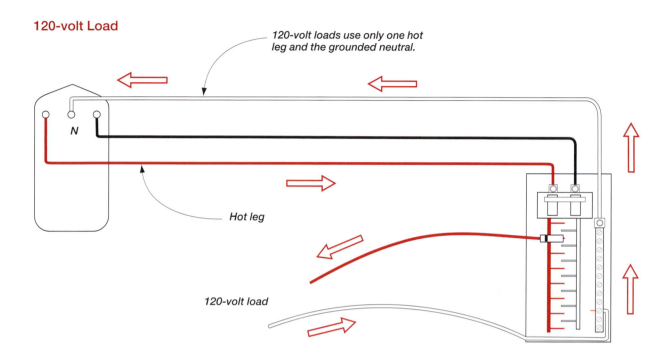

120-volt loads use only one hot leg and the grounded neutral.

Hot leg

120-volt load

120/240-volt Load

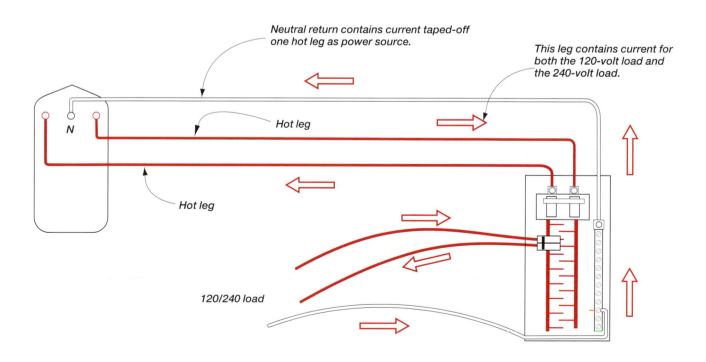

Neutral return contains current taped-off one hot leg as power source.

This leg contains current for both the 120-volt load and the 240-volt load.

Hot leg

Hot leg

120/240 load

the wire. The higher the material's resistance, the more energy it will take to strip off the electrons. The higher the resistance, the less current flow. The only way to increase the current in a circuit with a given resistance is to raise the voltage, which increases the electrical pressure.

Without any numbers, resistance is represented by the letter R. It is measured in ohms; a number quantity is always followed by the Greek letter omega (Ω) or the word "ohms" (30Ω or 30 ohms). Resistance is measured by a volt-ohm meter (VOM) or a multimeter, or it can be calculated if the voltage and current are known: $R = E/I$.

Resistance can also be the property of an appliance that converts electrical energy to heat energy. For example, let's take a common 240-volt, 4,500-watt water heater that's pulling 18.75 amps. As the 18.75 amps flow into the water-heater element, the element produces heat as it opposes the electron flow (the element has a resistance of 13 ohms). This opposition produces heat—exactly 4,500 watts worth—which turns the cold water hot.

POWER is the product of current and voltage. This is listed in watts or volt amps on all appliances. Volt amps come from the formula $P = \text{volts} \times \text{amps}$, or power equals voltage times current. Practically, we see power as the dissipation of heat, as in the water heater example previously mentioned. All appliances dissipate heat or provide energy; thus, you will always see the wattage listed on the appliance.

Common VOMs and multimeters cannot measure wattage (although commercial units can), but it can be calculated if two of the following are known: voltage, current, or amps.

1. If voltage and current are known, use $P \text{ (watts)} = EI$. For example, if E equals 120 volts and I equals 10 amps, the power is 1,200 watts (120×10).

2. If current and resistance are known, use $P = I^2 R$. For example, if I equals 10 amps and R equals 12 ohms, the power is 1,200 watts ($10^2 \times 12$).

3. If voltage and resistance are known, use $P = E^2/R$. For example, if E equals 120 volts and R equals 12 ohms, the power is 1,200 watts ($120^2/12$).

DC and AC

Electricity that flows in one direction is called direct current (DC). Some naturally occurring examples of DC are static electricity and lightning. Although DC generators do exist, DC is normally chemically generated, such as that from a battery. The current flows like a river from its negative terminal through a load (a light or other appliance) and returns to the positive terminal. DC is the power driving our cordless tools and appliances.

AC is the workhorse of the modern electrical industry. The voltage alternates back and forth in what is called a sine wave and, unlike DC, can be easily produced and distributed by generators. Thus, AC is what we use throughout our houses to run our plug-in appliances.

Although AC originates at power plants and is distributed around the country via high-voltage transmission lines, as far as the house is concerned, electrical power starts and stops at the utility transformer—specifically the three terminals on its side. The two end terminals are the ends of the secondary winding (the windings step down the utility voltage from several thousand volts to the 120/240 volts used in your home). The center terminal, or center tap, is connected into the winding halfway between the two (see the top drawing on p. 8).

The two outside terminals produce 240 volts. Current leaves one terminal, goes to the house (meter base, service panel, and load), and comes back to the

other terminal (then reverses). Unlike 120-volt circuits, 240-volt circuits don't use the neutral (center) tap. No neutral is required to operate 240-volt-only loads (see the bottom drawing on p. 8).

For 120-volt circuits, current leaves one terminal and comes back via the cable that connects to the center tap. The center tap is the neutral (see the top drawing on p. 9). The voltage from one terminal to the center tap is called one phase, or leg, of the incoming power.

The opposite transformer terminal also sends current into the house, and that current also comes back via the neutral to the center tap. This is the second phase, or leg, of the house power. Each phase represents a hot bus in the service panel.

For loads such as electric stoves and electric dryers that require both 120 and 240 volts, we have different load currents simultaneously flowing on both hot conductors and the neutral. Here, if the 120-volt current is "in phase" (that is, going in the same direction along the neutral as the 240-volt current), then the two currents in the neutral are added together. If the 120-volt current is "out of phase" (going in the opposite direction) with the 240-volt current, then the two currents are subtracted. This is why the neutral cable is allowed to be smaller than the hot cables (see the bottom drawing on p. 9).

Anatomy of a Residential Electrical System

The house wiring begins at the service entrance. In this section, I want to show you a typical house wiring system (see the drawing on pp. 12–13). I'll go into more detail about all these circuits in later chapters; this is just an example. It's also important to remember that

What "Neutral" Means

The neutral conductor is the conductor that leads back to the center tap of the utility transformer. It is connected to earth at two locations: the utility transformer and the main panel. That is why it is sometimes referred to as the grounded neutral conductor. Its designation is white or a white tape on a black cable.

Neutral current is commonly referred to as return current, because once the current passes through the load, the white insulated neutral conductor makes the complete circuit back to the utility transformer. All electrical current must make a complete loop. The current starts at the utility transformer and must therefore return to the transformer; the neutral is simply the return part of that loop.

The neutral cable or wire can kill you just as readily as a "hot" cable. The same current that flows in a hot conductor flows out the neutral. If you place yourself in series within that loop, even on the neutral side, you will be electrocuted.

different houses require different setups. I've already given you some background on where the power comes from. Now let's see how it all gets hooked up inside the house.

Let's go back to the service drop, or triplex (this example is an aerial feed). The utility wires splice onto the service-entrance wires next to the house at a drip loop. The utility's responsibility stops at the drip loop; from there, the owner's begins. The homeowner pays for problems on the house side of the drip loop. The utility pays for problems on the utility side. The masthead, which points down, and the drip loop, which circles up, keep water from following the wires, flowing into and down the mast, and entering the meter base.

Once the service wires enter the masthead (also called a weatherhead), they travel down the steel mast (or riser) into the meter base (not shown in the drawing). From there, the service-entrance (SE) cable goes directly into the main service panel. The main panel is where all wiring begins. From there,

The Main Wiring Runs in a Residence

A. Main service panel

B. "Home-run" cables from main panel to loads (run through attic or ceiling or through basement)

C. Cable for baseboard heater (dedicated circuit)

D. Utility-room receptacle (dedicated circuit)

E. Dedicated circuit for dryer receptacle

F. Cutoff switch (optional when in sight of panel) for water heater (dedicated circuit)

G. Range cable (dedicated circuit)

H. Range fan fed off living room circuit

I. Dishwasher cable (dedicated circuit)

J. GFCI-protected receptacles on countertop

K. Refrigerator cable (dedicated circuit optional)

L. Cable for dining-room receptacles

M. Cable for in-wall heater (dedicated circuit)

N. GFCI-protected receptacle and fan

O. Living room circuit

P. Kitchen overhead light powered off living room circuit

Q. Cable for smoke alarms and 3-way light switching

R. AFCI cable for bedrooms and bedroom smoke alarms

wires travel to various receptacles, lights, and appliances throughout the house. In addition, the main panel is where the house electrical system obtains its earth ground via the ground rods.

The drawing at left shows some of the main wiring runs: the utility room, the kitchen, the living and dining rooms, and the bathroom. These rooms are illustrated because each one has its own special requirements. Cables going through walls perform several different functions, but if needed, all of them can be routed together through the same holes in the studs. The cables in the utility room contain the 12-gauge utility-room cable (for the receptacle only), the four-conductor 10-gauge dryer cable, the 10-gauge cable for the water heater, and the 12-gauge cable for the baseboard heater. Armored cable (see p. 23) is used for the exposed run from the water-heater cutoff box to the heater. In the utility room, there is only one 120-volt receptacle—for the washing machine. Most utility rooms require more than one receptacle. Lighting for the utility room (not shown) is pulled off of the living room circuit.

The bathroom receptacle is powered off its own 12-gauge circuit. Ground-fault circuit interrupters (GFCIs) protect all of the outlets; the in-wall heater has its own dedicated circuit.

All habitable rooms must have switched lighting at the point of entry. Receptacles and switches in the living and dining rooms are powered off of individual circuits for each room. The cable enters through the ceiling and connects to a receptacle, which feeds other receptacles and switches in the room.

The kitchen circuits include two 12-gauge small-appliance circuits, the lighting circuit, the 6-gauge (four-conductor) stove cable, the 12-gauge dishwasher cable, and the 15-amp refrigerator circuit, as required by code.

■ WORK SAFE
■ WORK SMART
■ THINKING AHEAD

For easy NM cable gauge recognition, cables should have color-coded jackets. Unfortunately, when I asked one manufacturer to make them this way, he refused. To solve this problem, I paint one side (paint only the side) of the NM cable a different color for each gauge (see chapter 6). Being able to tell a cable's gauge at a glance—even from across the room—will aid both you and your electrical inspector.

Receptacles along the countertop are GFCI-protected.

Not shown on the drawing are the three-way switching circuits and the smoke-alarm circuits that I would put on the 14-gauge circuit. Also not shown are the Arc-Fault Circuit Interrupter (AFCI) circuits that feed the bedroom outlets (lights, receptacles, and smoke alarms, if any).

Wire Gauges

We use insulated wire or cable to deliver electricity to where we need it. It didn't take long to learn that the larger the wire's diameter, the more electricity we could deliver to the load and the less we would waste along the wire. The American Wire Gauge (AWG) provides numbers for the different wire diameters. For some reason lost to history, the larger numbers are used to represent the smaller wire diameters. For example, a cable that has a big number, such as 18, has a smaller diameter than a cable with a small number, such as 1. Zero is called ought; a cable with one zero is written as 1/0—or one ought. And the system doesn't stop at zero for the largest diameter cable, either. No, that would be too simple. Numbers less than zero identify bigger cables. What's less than zero? More zeros. From 1/0 it goes to 00 (2/0), 000 (3/0), and 0000 (4/0). Each descending number indicates a larger diameter cable. To make things even more complicated, the system changes in midstream. As soon as you get past 4/0 (four ought), the diameter of the conductor is referenced by its cross-sectional area: 250 kcmil, 300 kcmil, and so on.

One of the biggest problems you will have in the field will be differentiating cable and gauges. It's sometimes hard to see a difference between cables with gauges that are only one or two sizes apart. Theoretically, the gauge number is supposed to be written on the cable's jacket, or sheath. Unfortunately, many cable manufacturers don't use paint; instead, they emboss the gauge number on the jacket. Embossed gauge numbers are almost impossible to see; consequently, many people wind up installing the wrong cable for the job. In my hometown, our new library had all of its 12-gauge and 14-gauge wires intermixed when it was built. The inspector didn't catch it because he couldn't read the wires.

What gauge goes where

Service-entrance cable for dwellings has its own reference in the NEC (see chapter 3). The ever-popular Table 310-16 gauge-finder is fine for everything except the service-entrance and the subpanel feeder cables.

ABOVE CODE With few exceptions, Table 310-16 of the NEC dictates the gauge size for the rest of the house. In practice, it breaks down to using 18 and 16 gauge for low-voltage circuits, such as doorbells and thermostats; 14 or 12 gauge for receptacles and lights; 12 gauge for most hardwired appliances; 10 gauge for electric dryers and electric water heaters; and 6 gauge copper for electric stoves. Code makes you install 12 gauge in the kitchen/dining/pantry area, as well as in bathrooms and utility areas. All other areas are allowed to be either 12 or 14 gauge.

The problem with this allowance is that 14-gauge wiring isn't always heavy enough. It also creates the potential for intermixing 14- with 12-gauge wiring. Instead, I use 12 gauge as much as possible and 14 gauge only for the optional refrigerator circuit, three-way light

Cross Sections of Copper Conductors

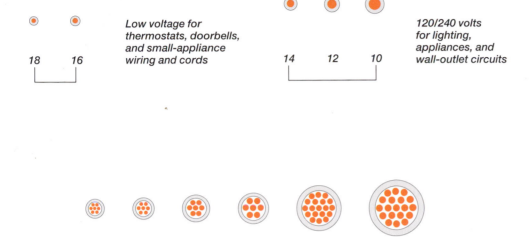

Low voltage for
thermostats, doorbells,
and small-appliance
wiring and cords

18   16

120/240 volts
for lighting,
appliances, and
wall-outlet circuits

14   12   10

8   6   4   2   1/0   2/0

240 volts for major-appliance circuits, service entrance, and subfeeds

**The 6-gauge stove wire can
carry 2½ times more current
than the smaller 12 gauge.**

Deciphering Cable and Wire Codes

The only place you'll find more codes than on electrical conductors is in a spy novel. The gauge and manufacturer are not the only things stamped on the outside of a conductor.

A multidigit code indicates at what temperature range the conductor can physically be installed. You don't want to put a conductor that's approved only for dry locations in a wet area (for example, underground), and you don't want to put a conductor approved only for low temperatures in an area with high temperatures (such as an attic). Not only do you have to figure out the multidigit code, but you also must watch for preempt codes, which change or alter the previous code letters, depending on the installation.

The letter codes that appear after the gauge number and manufacturer's data refer to the insulation of the conductors. The letters tell you the type of insulation material, which is normally thermoplastic; the maximum temperature the insulation can be subjected to without damage—60°C, 75°C, or 90°C—and whether the conductor can be installed in wet locations, dry locations, or both.

Because the manufacturer is trying to tell us only three things, you'd think it would be rather simple, but it's not. In addition to the basic coding and the preempt coding, it is also significant if something is missing (I'm not kidding).

NM and UF Cables

Here's a sampling of the coding that appears on cables and other conductors commonly used in a residence:
H: If present, the maximum allowable temperature is 75°C. If not present, the maximum temperature allowed is 60°C. NM cable uses the letter B instead of H, but it works the same way: with B, 90°C; without B, 60°C.
HH: The maximum temperature is 90°C.
N: Nylon sheath around thermoplastic insulation.
T: Thermoplastic insulation around wire.
W: If present, the conductor can be used in wet and dry locations. If not present, it can only be used in dry locations (sometimes damp). If the letter W appears after HH—W is a preempt coding—the insulation's maximum temperature has been reduced to 75°C for wet locations (it is still 90°C for dry locations).
X: Cross-linked polyethylene insulation (normally used for SE cable).
–2: A preempt suffix that means the conductor can be used at temperatures up to 90°C, wet or dry.

Wire gauge	"With ground"	Nonmetallic sheath		Insulation rating: 90°C	Maximum voltage allowed	
10 – 2	W/G	TYPE	NM	B	600V	
12 – 3	W/G	TYPE	NM	B	600V	
14 – 2	W/G	TYPE	UF	B	600V	Sunlight resistant

Number of conductors in cable · Underground feeder · Cable protected against sunlight

SE Cable

Though NM cable is fairly simple to decipher, SE cable is another story. For example, if SE cable has XHHW-2 stamped on it, the insulation is rated up to 90°C. However, the W preempts the HH and indicates that the conductor can be used in wet locations down to 75°C. The –2 preempts the W and gives a 90°C rating to both wet and dry locations. (Confused yet?)

Aluminum   Style of cable   Number of hot conductor; in this case, two insulated 4/0 conductors.   Gauge of third conductor. XXHW means the third conductor is bare and braided.

AL TYPE   SE CABLE STYLE U XHHW-2   600V 2CDR(4/0)   1CDR(2/0) (UL) 1996

X refers to the cross-linked polyethylene jacket. HH means the temperature rating of conductors is 90°C. W preempts HH and lowers rating for wet locations to 75°C. –2 preempts W and raises the temperature rating back to 90°C.

Voltage rating   UL listed   Year manufactured

THHN and THWN Wires

Using the codes, let's decipher THHN and THWN conductors. THHN is sometimes used in a home whenever conduit is required. It has thermoplastic-insulated wire with an outer nylon sheath and can be used in dry areas where the temperature could reach 90°C. THWN, also common in homes and run in conduit, has thermoplastic-insulated wires with a nylon sheath and a temperature rating of 75°C in both wet and dry locations.

Wire gauge   Thermoplastic insulation   90°C   Nylon sheath

12AWG   600V   T   HH   N

12AWG   600V   T   H   W   N

Maximum voltage rating of wire   75°C   Wet and dry locations circuits, and the smoke alarms. Keeping these circuits at 14 gauge saves me from having to use expensive 12-gauge three-way switch cable for loads that pull little or no current.

Copper vs. aluminum wiring

When it comes to residential wiring, there are two standard wire materials: aluminum and copper. Most homes use copper wiring for receptacles and switches for good reason. Small aluminum wires have to be handled very carefully because they break easily. And they must be installed and spliced only with code-approved materials. Sometimes those materials are hard to find. For the larger-diameter wires (6 gauge and up), electricians tend to use both copper and aluminum. Aluminum has a slight cost advantage over copper, and this is more apparent in the larger-diameter cables.

When exposed to air, the surface of bare aluminum oxidizes. On one extreme is minor oxidation, which is a dull gray. On the other extreme is major oxidation, which creates a white powder that can grow like a tumor and bind everything together tighter than a thread-lock compound. The latter is a more serious problem, but even minor oxidation insulates the strands of aluminum from each other, as well as from the connector, increasing the resistance and degrading the metal. Therefore, whenever you strip insulation off aluminum, you must apply an anticorrosion coating to the bare wire—not just on the part connecting to the terminal—to prevent oxidation. I also recommend wire brushing the aluminum to remove any oxide on the surface

Preparing aluminum cable (left to right). Remove the insulation to expose the conductor. Next, wire brush the conductor until it's shiny. Finally, use your fingers or an old toothbrush to cover the bare conductor with an anticorrosion compound.

before applying the antioxidation compound.

Splicing aluminum wire

It's easy to splice large-diameter aluminum wire with one of the dozens of devices sold everywhere for that purpose. I crimp wire together with specialized, expensive, long-handled crimpers. You can even splice service-entrance conductor (SEC) with this tool because it is a nonreversible device. For non-SEC large-diameter splicing jobs, you can use crimpers or large lugs and large split-bolt clamps. Bimetal tongue split-bolt clamps can splice large-diameter copper cables to large-diameter aluminum cables. There is even a special toothed-type fitting that tightens with a large socket and ratchet and cuts through the insulation to make a splice.

What about small-diameter, single-strand 10- and 12-gauge aluminum that was once used on 15- and 20-amp branch circuits? That experiment failed many years ago; however, most manufacturers have largely ignored the problem. I developed several ways to deal with aluminum circuits. One was cutting up neutral buses. The bus is approved for both aluminum and copper wires, so it seemed like a logical adaptation; I modified it to fit into a smaller outlet box. Today, Ideal, Burndy, and several other companies have such a device on the market. The biggest problem with the device is that it takes up too much room in the outlet box.

ABOVE CODE The best way to connect a single-strand aluminum wire to a single-strand copper wire is with a simple split-bolt clamp with tongue. The Madison #13-10 is approved for copper-to-copper, copper-to-aluminum, and aluminum-to-aluminum connections. Amp makes a crimp splice but it

is entirely too expensive—around $50 to $100 apiece. Ideal Industries, Inc. makes a purple wire nut that will make the splice but, allegedly, there have been some installation-related problems. I prefer using the tiny split-bolt clamp. Once you've made the splice, however, you must insulate any bare metal with three layers of high-quality electrical tape.

The Wire We Use

New houses use dozens of types of wire and cable. To make things even more confusing, wiring systems have evolved over the years. Some older homes may have a combination of several different wiring systems. Following is a brief overview of what you might find in your home.

With renovations, you need to make doubly sure that the wiring has been done correctly. I once inspected a house that had all its old wiring replaced. The man selling the place even bragged about it. Upon testing the receptacles, I discovered that not one indicated the presence of a ground wire. To find out what the problem was, I inspected several receptacles and, sure enough, there was new NM wire. But the ground wire on each cable was cut back to the jacket. Apparently, the seller hired a couple of old-timers to do the work . . . and they didn't know what to do with those "confounded bare wire things."

Knob and tube

The oldest wiring that you might see in a house is knob and tube. The wiring system gets its name from the porcelain knobs and tubes that the wire runs in and around. The wiring is still legal to use and can even be installed for new

---

■ WORK SAFE
■ **WORK SMART**
■ THINKING AHEAD

Use copper cable instead of aluminum cable whenever possible. Copper has fewer corrosion problems and is a better conductor (especially over long distances), which means that you can use a smaller-gauge wire for the same amperage circuit. You'll find that the smaller-gauge wire is easier to run and takes up less space in the service panels and conduit.

Splicing Single-Strand Aluminum to Single-Strand Copper

This rectangular butt connector can splice both copper and aluminum (14 gauge through 6 gauge). To use, insert the copper wire on one end and the aluminum on the other, and then tighten down the screws.

The split-bolt connector has a center tongue that separates the two wires being spliced so they don't touch each other. This is what allows the connector to splice different metal wires, such as those of aluminum and copper.

SPLIT BOLT CONNECTOR

QUANTITY
25

CATALOG NUMBER
ASB-10

SIZE
#16-10

USES:
ELECTRO-TIN PLATED SILICON BRONZE, FOR COPPER TO COPPER, ALUMINUM TO COPPER, AND ALUMINUM TO ALUMINUM CONNECTIONS

MADISON
Cleveland, Ohio 44103

This tiny split bolt is my preferred method of splicing single-strand copper and aluminum together. As you can see from the box, it is approved for 16 gauge through 10 gauge—perfect for splicing 10-gauge aluminum to 12-gauge copper.

extensions off an existing knob-and-tube system, assuming you can find the material. Knob and tube consists of two parallel running wires (no ground wire). The copper wires have a rubber coating and a single-cloth braid—black for the hot wire and white for the neutral or return (most wire is so old and dirty that both colors look the same).

Knob and tube doesn't need to be replaced just because it's old. As long as the insulation is intact and you don't need a ground connection on the receptacles it feeds, there is no reason to replace the wires. But you do need to check a couple of things. Look for any sections of bare wire—spots where the insulation has hardened, cracked, and broken away. Sometimes this happens from old age, other times from overfusing and overheating the wire. If you move into a house that has overfused 14- and 12-gauge circuits (normally at 30 amps) be especially wary about the wiring. To be safe, keep the old wiring fused at 15 amps even if it is 12 gauge. If you find any bare spots where the insulation has broken away, have the house rewired. Some knob-and-tube systems had a two-fuse protection system: one for the hot wire and one for the neutral. You are no longer allowed to fuse the neutral, so if you have this system, you'll need to wire around the neutral fuses or replace the wiring entirely.

Another problem with knob-and-tube wiring is that it was designed for open-air heat dispersion. That is, it was never designed to be covered with insulation. The problem is exacerbated if the house has blown-in insulation in the walls, because the knob and tube in the walls can no longer dissipate the heat. So, what do you do? You can continue to use the wire for light-duty circuits but fuse it at 15 amps, as previously mentioned. If the gauge is already 14, I would

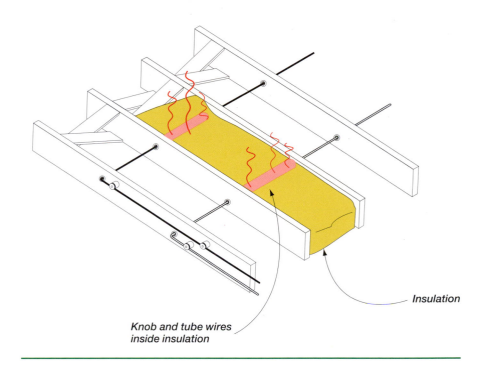

Insulation Causes Overheating

Knob and tube wires inside insulation

Insulation

not advise using it at all, except for very light-duty loads.

Grounding is another problem. Kitchen, bath, and utility room circuits need a ground, as do specific appliances, such as a computer. Thus you'll need new circuits for those loads. The NEC does not require splice boxes to be used when splicing knob and tube. It does state that you should solder the splice as they did back in the old days or use an approved connector. Practically speaking, soldering isn't done anymore, so a noncode splice is almost guaranteed. If you see such a splice with modern plastic electrical tape, the NM wire probably isn't soldered. If you see the old rubber tape, it is probably soldered per code.

The best way to splice into knob and tube is a moot point, because you cannot get a ground from it and the old circuit shouldn't be loaded any more than it already is. If you need to add on to an old knob-and-tube circuit, run a new cable back to the service panel. For more infor-

Round Solid Porcelain Knobs

63N3925—Old Code No. 5½ solid porcelain knob. Height, 1⁹⁄₁₆ inches. Diameter, 1 inch. Hole, ¼ inch. Groove, ⁵⁄₁₆ inch. Weight, per 100, 8½ pounds. Price, each, **1 c**; per 100, **92 c**; per 1,000..........................**$9.00**

63N3927—New Code No. 5½ solid porcelain knob. Height, 1⁹⁄₁₆ inches. Diameter, 1⅛ inches. Hole, ¼ inch. Groove, ⁵⁄₁₆ inch. Weight, per 100, 11½ pounds. Price, each, **1½ c**; per 100, **$1.18**; per 1,000...**$11.20**

63N3929—No. 4 solid porcelain knob. Height, 1¹¹⁄₁₆ inches. Diameter, 1½ inches. Hole, ⅜ inch. Groove, ⅜ inch. Weight, per 100, 20 pounds. Price, each, **2 c**; per 100, **$1.57**; per 1,000....**$15.10**

63N3931—No. 4½ solid porcelain knob. Height, 1⅞ inches. Diameter, 1½ inches. Hole, ⅜ inch. Groove, ⁷⁄₁₆ inch. Weight, per 100, 21½ pounds. Price, each, **2¼ c**; per 100, **$1.76**; per 1,000....**$17.00**

Round Split Porcelain Knobs

63N3935—Old Code No. 5½ porcelain split knob. Height, 1¾ inches. Diameter, 1 inch. Hole, ¼ inch. Grooved to take two No. 12 or 14 wires. Weight, per 100, 9½ pounds. Price, each, **2 c**; per 100, **$1.56**; per 1,000 **$15.00**

63N3937—New Code No. 5½ porcelain split knob. Height, 1¾ inches. Diameter, 1⅛ inches. Hole, ¼ inch. Weight, per 100, 11 pounds. Price, each, **2¼ c**; per 100, **$1.75**; per 1,000.**$16.90**

Porcelain Tubes

Unglazed Porcelain Tubes, ⁵⁄₁₆ inside; ⁹⁄₁₆ outside. Take either 14, 12 or 10 single braid rubber-covered and weather-proof wire. Required wherever a wire is drawn through a partition or joist of any kind. Length given is from underhead to end.

Article Number	Length, Inches	Each	Per 100	Per 1,000	Weight per 100 lbs.
63N3902	3	1 c	$0.78	$ 7.15	7½
63N3904	4	1½ c	1.25	11.95	9
63N3906	6	2 c	1.73	16.80	13
63N3908	8	4 c	3.30	31.35	15½

Porcelain Cleats

Glazed Porcelain Cleats. Take either 14, 12 or 10 wires. Two-wire have 2½-inch wire centers. Three-wire have 1¼-inch wire centers. Weight, per 100 pairs, 25 pounds.

63N3920 — Two-wire.
Price, per pair......................................**$0.02¼**
Per 100 pairs.....................................1.95

63N3922—Three-wire. Price, per pair............**.02½**
Per 100 pairs.....................................2.20

These photos, one from an early Montgomery Ward catalog and one from a Sears, Roebuck and Co. catalog, show wire and cable from 1900 to 1920. The knobs, tubes, and cleats were used for knob-and-tube wiring systems. Note that duplex cable was around even then. (Photos by Scott Phillips.)

New Code, Rubber-Covered Copper Wire, Single Braid

Solid conductor wire, insulated with rubber compound over which is one cotton saturated braid. Recommended for any open wiring on cleats, tubes or knobs, and for loom and moulding wiring.

Each unbroken coil bears Underwriters' Inspection tag. Full coils contain 100 feet or 500 feet, or we will cut wire to any desired length.

Article Number	Size	Weight per 100 ft Pounds	Per Foot	Per 100 Feet	Per 1000 Feet
63N3015	14	3	1½ c	$1.40	$10.25
63N3020	12	4½	2½ c	2.20	19.75
63N3025	10	6	3¼ c	3.00	27.40

New Code, Rubber-Covered Wire, Double Braid

Offers better protection than single braid wire. Should be used in all metallic conduits, flexible or rigid. Has one solid conductor, insulated with rubber compound over which are two saturated cotton braids.

Article Number	Size	Weight per 100 ft. lbs.	Per Foot	Per 100 Feet	Per 1000 Feet
63N3040	14	4	2 c	$1.70	$16.20
63N3045	12	5½	2¾ c	2.35	22.25
63N3050	10	7	3½ c	3.20	29.90
63N3055	8	10	5 c	4.55	43.95
63N3060	6	13	7½ c	7.05	67.50
63N3065	4	18½	10½ c	9.90	96.25

Duplex New Code Rubber-Covered Wire

Consists of two solid conductors, each insulated with rubber compound, over which is one saturated cotton braid. Conductors so insulated are laid parallel and covered over all with saturated cotton braid. Convenient for wiring in metallic and non-metallic conduits.

Article Number	Size	Per Foot	Per 100 Feet	Per 1000 Feet
63N3080	14	3¼ c	$3.00	$27.50
63N3085	12	5 c	4.75	44.95
63N3090	10	7 c	6.60	65.25

New Code Electric Fixture Wire

For wiring Electric Chandeliers, fixtures, brackets, lamps, etc. Also used for short lighting circuits. Solid copper conductors. No. 18 size, has ¹⁄₆₄-inch, new code rubber compound covering. No. 16 size, ¹⁄₃₂-inch new code rubber compound covering. Each is covered with saturated cotton braiding.

Article Number	Size	Per Foot	Per 100 Feet	Per 1000 Feet
63N3102	18	1 c	$0.83	$ 7.70
63N3107	16	1½ c	1.20	10.45

Weather-Proof Copper Wire

For use on outside work only. Not to be used indoors. Withstands weather better than rubber-covered wire. Conductor is of solid copper wire, covered with a triple braid of weather-proof insulating material.

Article Number	Size	Wt. per 100 ft. Pounds	Per Foot	Per 100 Feet	Per 1000 Feet
63N3125	14	2½	1½ c	$1.15	$ 9.95
63N3130	12	3½	1¾ c	1.55	14.50
63N3135	10	5¼	2½ c	2.30	21.25
63N3140	8	7½	3½ c	3.15	29.25
63N3145	6	11	5 c	4.70	43.65
63N3150	4	16½	7½ c	6.90	63.90

Knob-and-Tube Splice

This is the typical method of splicing into knob and tube. However, there will be no ground on the tap cable.

Junction box

Add three layers of tape over knob-and-tube wire where it goes through the NM connector.

Connecting NM cable

Tap cable

mation on knob-and-tube wiring, read Article 324 of the NEC.

Armored cable (AC)

Perhaps just as old, and sometimes running alongside knob-and-tube wiring, is a cable system we generally call armored cable (AC) or Greenfield (from Harry Greenfield, who co-invented it). This cable came in many strange versions but evolved into the now-familiar spiraled metal (normally steel). You can buy the armor separate or with cable already inside. It was extremely popular from 1910 to 1930. BX is the trademarked colloquial term that comes from the very popular GE product.

Aluminum AC came out in 1959, as did the requirement of having a bonding strip, or wire, run the complete length of the armor but electrically connected to nothing. The bonding strip's purpose was to lower the overall impedance of the armor. The basic problem with AC was that the long, continuous spiral of metal produced tremendous inductive impedance, a special type of resistance associated with coiled wire. Therefore, when working with old AC, do not remove the bonding strip (assuming it had one—

Armored cable terminates in an outlet box in a 1930s home.

Even close up you can just barely differentiate the black (hot) from the white (neutral) conductor. The insulation on both wires is severely deteriorated and may break away if the wires are moved.

some versions didn't) or tie it into the grounding system. Bend it over the end of the AC as if to hold the anti-short bushing in place. More often than not—especially in short runs—the strip will be missing.

AC cable was first put in homes that had ungrounded receptacles, because grounding wasn't thought to be that important. When that assumption was proven wrong, the steel spiral was used as the grounding circuit in three-prong grounding receptacles. My county seat used ungrounded AC until the introduction of computers. Using the steel AC as the grounding circuit has kept things working fine to this day. If you have this ungrounded system in your home and the armor is intact, you can do the same. If you see that the conductor's rubber insulation is cracking and breaking apart within the receptacle and switch boxes, it would be a good idea to rewire the entire house. For repairing short runs, pull the old wire out of the AC and slide in new wire—normally THHN/THWN or NMB. To cut into the armor, you need a special tool (see the sidebar on the facing page). You can use a hacksaw, but it is almost impossible to cut into the cable without damaging the conductors.

Metal-clad cable (MC)

Metal-clad (MC) cable has taken the place of AC cable. MC cable looks a lot like AC cable, but within its metal jacket it has a green insulated ground wire (the metal jacket cannot be used for a ground). You can use this wire system in lieu of NM cable throughout the house but, because of code changes, it is more commonly used in commercial work. However, if you have a multistory house, you may come across it on your upper floors. (At one time, code limited the number of floors on which a residence could use NM cable.) Code no

longer prohibits NM cable in a residence, regardless of height.

MC cable must be kept dry, but it can be embedded in plaster and run through block. It also must be supported within 12 in. of the outlet box and no farther than 6 ft. apart after that. Its advantage over NM cable is that rats and mice cannot gnaw on it. Its biggest disadvantage is that it is very labor-intensive to install, and you need to use special bushings to prevent the metal's sharp edges from cutting into the wires. (If you are planning to run rigid conduit, MC cable can be a lot faster to install than NM cable.) You also need a special tool to cut into the jacket, special metal boxes to terminate the cable, and special connectors.

My advice? Don't use MC cable in a residence unless you have a specific reason to do so. Once I had a customer insist on MC cable because she owned an old house and thought that mice would chew on the wires. When she saw how problematic (and costly) MC cable was to install, she eventually changed her mind. As for the mice, her cats took care of that problem.

Nonmetallic sheath cable (NM)

NM cable consists of two or more insulated conductors in a single nonmetallic jacket (two insulated conductors in an NM jacket is called a duplex). It is not a modern invention. Duplex cable has been around since before 1918. Early duplex NM cable consisted of two insulated conductors in a silver-colored jacket or a braided, brown-jacketed cable. The silver-colored jacket cable was used to wire houses in the early '50s; the brown-jacketed cable was used to wire houses in my section of Virginia in the '20s and '30s.

Modern NM cable (also called Romex®, a registered trademark of the Southwire Company) came into

Cutting MC and AC Cable

Insert cable in the long slot down the center of the tool. Position the cut line over the cutting wheel. Tighten the knurled knob on top of the tool until the cable is wedged tightly in the slot.

Turn the handle to engage the cutting wheel. The wheel will slice through the metal but not the wires.

Remove the cable from the tool, and then trim back and remove any remaining cut spirals.

Insert plastic anti-short bushing between the metal and the plastic sheath (the bushing here is orange, but the material comes in different colors for different sizes).

Avoid buying NMB cable if you cannot read the writing on the sheath. The printing on the embossed cable (top) and painted cable (middle) are both illegible. Only the Essex cable (bottom) is legible.

NMB cable is made up of individual, thermoplastic-insulated conductors, each slipped into a clear nylon sleeve, then wrapped in paper insulation and a thermoplastic jacket. The clear sleeve gives NMB cable a higher temperature rating than the older-style NM cable.

Old NM cable is made up of individually insulated conductors, lots of paper insulation, and a hidden zip cord to rip the cable, all contained inside a woven sheath.

use around 1965. There are four basic types of NM cable: old-style NM, NMC, NMB, and UF. Each one has a specific job. Old-style NM was the most common cable used in home wiring until a few years ago. It is now replaced by NMB cable because of the need for a higher insulation temperature. Old-style NM is rated at 60°C; NMB is rated at 90°C.

Generally, NM-style cable has two or more thermoplastic insulated wires, one bare grounding wire, paper insulation surrounding the conductors, and a thermoplastic jacket or sheath. NMB cable is the NM style that is currently used around the home for receptacles, lighting, and small-appliance circuits. Modern NMB cable can be used for both exposed work (as long as it is not in physical danger) and concealed work in dry locations, but it cannot be embedded in masonry, concrete, adobe, dirt, or plaster. NMC cable is the same as NMB cable, but its outer jacket is corrosion resistant.

Type UF (underground feeder) cable looks like NM cable, but UF has the grounding wire, the neutral wire, and the hot wire embedded in solid thermoplastic. UF cable is used in wet locations, such as an underground circuit to an outdoor light, where NM cable would not be allowed by code. All NM cable, even UF, is normally sold in 250-ft. rolls. Some stores that cater to do-it-yourselfers sell the cable in shorter lengths; however, the markup may be considerable.

Service-entrance cable (SE)

Type SE (service entrance) cable brings power into the house from an overhead splice at the utility cables. SE cable comes in different styles: U, R, and USE. Style U is flat and approved only for above-ground use. This is the cable you see attached to the outside of most houses. It has two black insulated wires (sometimes one has

Use UF (underground feeder) cable for wet and buried locations.

a stripe for phase identification) surrounded by many strands of braided bare wire for the neutral and is covered by a layer of glass-reinforced tape and protected by a PVC jacket. The jacket is flame-retardant, moisture-resistant, and almost entirely UV-resistant.

Style R is round and has three insulated conductors with one bare grounding wire covered with a layer of glass-reinforced tape and protected by a PVC jacket. Like style U, R is approved only for above-ground use.

Individual insulated conductors rated USE (underground service entrance) are normally used when the service must be located underground or within conduit. USE conductors are available in gauges of 6 and higher.

SE cable can also be used to provide power to some appliances, such as 240-volt-only heat pumps and electric furnaces. However, it cannot be used where an insulated neutral is required. SEU cable cannot be used to feed 120/240 appliances, such as electric dryers and stoves. Such appliances need three insulated conductors plus ground (two insulated conductors for the 240-volt load and a return neutral for the 120-volt load), and standard SE cable has but two. However, SER cable can be used to feed 120/240-volt loads (subpanels and heavy-current loads), because it has three insulated conductors with an additional bare grounding conductor. SE cable is covered in detail in article 338 of the NEC.

THHN/THWN is also used for service-entrance cable, but it has to be in conduit. If you want to run the cable out of the weatherhead for an aerial drip loop, contact your inspector to verify that it is allowed.

Type-U cable with XHHW insulated conductors is perhaps the most common service-entrance cable used. It has thermoplastic insulation, glass-reinforced tape, a braided neutral, and two insulated conductors.

THHN/THWN is the most common SE cable used in conduit. Smaller gauges can be used as branch circuit wires in the house as long as they stay in conduit.

The two most common types of rigid conduit used are PVC (plastic) and EMT (electrical metallic tubing).

Conduit Systems

Conduit systems provide physical protection to cable or wires. The rule is that if the conductors have a jacket around them as a group, then that is their physical protection. If the conductors do not, and if they are individual conductors, such as THHN/THWN, then they have to be run in conduit. In addition, if the cable is open to significant abuse, then the cable or wire must be in conduit.

The two most common conduit systems for branch circuit wiring are plastic (PVC) and electrical metallic tubing (EMT). Both types of conduit can be used indoors or outdoors. PVC is glued together and has the advantage of being watertight and not needing to be grounded. A common application of both types is in a basement or garage that has finished block walls. You would normally attach a piece of treated wood onto the wall and mount the conduit and boxes to it. This is much easier than trying to mount it on a block wall, unless you have a hammer drill that can drill into the cement easily. The wiring inside the conduit can be NM or THHN/THWN. If you have a complete conduit system, use the latter. THHN/THWN is cheaper and faster to install and it takes up less room in the conduit. However, most people just use the vertical sections. They'll run NM cable overhead and then slip it into the vertical conduit as it goes down to the receptacle and switch circuits. This gives you the advantage of running the flexible NM cable overhead and the physical protection where it is needed.

A newer-style conduit system that is gaining in popularity is called moisture tight. Commonly available at most supply houses, it is very easy to install. This nonmetallic, nonconductive, noncorrosive conduit is extremely flexible and can be used to protect wire/cable going to heat

Moisture-tight flexible conduit is very popular for running wire and cable outside. It can be bought in rolls or by the foot and is available in various diameters.

The most common error made with PVC and EMT conduit is using too small a diameter when trying to slip in NM cable. Small (½-in.) conduit allows you only enough room to slip in one cable. Comparatively, ¾-in. and 1-in. conduits allow for a lot more room and freedom of design. Just make sure you pick boxes that are prepunched for those two sizes.

Moisture-tight fittings seal on both sides. The threaded end fits into a box knockout and is sealed by an O ring. The opposite end screws into the conduit.

The finished product: The fitting is attached to the conduit.

Forget using a hacksaw to cut large-diameter cable (XHHW); instead, use a reciprocating saw with a fine-tooth blade.

duit installs like a garden hose; all you need is an adjustable wrench.

Running groups of cables

Running a large number of cables through joists or across the top of attic beams can get you into trouble. On one job (which failed inspection) a man had all the cables bundled tightly together. It looked beautiful—15 to 20 cables all wrapped together going through the joists as neat as a pin. But such beauty has a major engineering drawback. The cables on the inside of the bundle will overheat because they cannot get any moving air. The heat will break down the insulation, which will eventually lead to a short. To avoid this problem, always spread the cables through several holes when running through joists or studs. When running the cables on top, spread them flat over a large area. Locate attic cables to one side—choose a side where there is little headroom and fewer chances of someone tripping over them.

pumps, air conditioner hook-ups, spas, and anything else that has outside wiring. PVC can be used in its place, but it is harder to install and cannot take vibration. (Because of this, inspectors sometimes fail heat pump installations where PVC conduit is used.) Moisture-tight con-

A common scissors-type cutter works well for cutting cable and individual conductors.

CHAPTER 2

Tools of the Trade

ELECTRICAL TOOLS haven't changed that much since I wrote the first edition of this book. My everyday tools are pretty much the same. I still use the same broad-nosed electrician's pliers to cut cable and pull wires. I still use a needle-nose pliers to bend stripped wire ends into half-circles to put under screws. However, old dogs can learn new tricks, especially when it makes work more comfortable and efficient. These changes are included along with the basics.

If you are a first-time reader, pay close attention to my favorite tools, such as my Klein electrician's pliers and my Gardner Bender (GB) automatic wire strippers. Other tools, such as the common cable ripper, are notably absent. The problem with rippers is that as they dull, they tend to bunch the sheath up around the cable, forcing me to use my utility knife. Rippers are also hard to use in tight locations, like a crowded outlet box. And even if you get the sheath ripped, you'll still have to use a knife to cut it away from the cable. I've found that using a utility knife is

equally safe and easy (see Stripping Cable Safely on p. 37).

Buying quality tools is important, and so is maintaining them. You can judge how serious people are about their

The little red wagon that I "borrowed" from my granddaughter works perfectly as a carryall and home base for all my job-site tools.

Insulated and Grounded Tools

A few years back, my wife was working with me on a wiring job. Before cutting a particular wire, she asked if it was dead. "Yes," I yelled back. The instant she cut the wire, there was a large "crack," her pliers started spitting sparks and, finally, the breaker kicked. "Whoops," I said.

The moral to this story is that accidents happen. For that reason, every tool you use needs to be insulated or grounded. Both types of protection will save your life if you accidentally cut a hot wire.

Most hand tools can't be grounded (although some electronic hand tools do have ground leads). Instead, look for good-quality insulated handles. If the plastic starts slipping off or appears cracked, repair or replace the tool immediately.

Not all plastic handles are designed to provide electrical protection. To be safe, look for a double overlapping triangle with "1000" stamped immediately to the right. This symbol means that the handle is designed to protect users up to 1,000 volts. In addition, these tools have an integral handle guard that prevents your hand from slipping down the tool and onto a live wire.

AC power tools need to be double insulated or grounded. Double insulated means all live electrical parts are double insulated from the user. If an internal short occurs, the tool will direct the current back to the panel and trip the breaker. A grounded tool has all metal surfaces connected to the ground wire in the cord. In this case, you must use a three-prong grounded receptacle to use the tool safely.

Safe

Double insulated

Two-wire cord, no ground

Hot-neutral short

Not Safe

All-metal frame

Hot frame

Current flowing through the tool user

Two-wire cord, no ground

Hot-to-frame short

Two-wire W/G cord

Hot-to-frame short

Fault current on equipment grounding conductor going back to the panel

All-metal frame

Hot frame

Current flowing through the tool user

Cheater plug

Hot-to-frame short

Two-wire wall outlet

work by the quality and maintenance of their tools. It's equally important to use the right tool when working with electricity. One of the first lessons an electrician learns is that the right tools can make a hard job easy and safe—and the wrong tools can make an easy job hard and dangerous. (To borrow from an old phrase: There are old electricians and there are foolish electricians, but there are no old, foolish electricians.)

In this chapter, I will tell you what I've learned about electrical tools from my many years of using (and abusing) them. And I will throw in a few tricks of the trade as well. I'll show you the tools an electrician uses every day, including how to use them safely and how to maintain them to ensure a long life span—for both tool and user. And here's an important note: When on the job site, many tools look the same, and someone may claim yours. To help prevent this, put your name on all your tools with paint, stickers, or other means. My favorite method is paint. If you paint your tool a solid color, you will be able to recognize it from across the room—especially if it's in someone else's hands.

General Tools

General tools are the tools you'll be using most often. Although I can't cover every tool, I will discuss the ones that are most important to me, how to use them, and their advantages and disadvantages. Stay away from dime-store specials. Quality tools may be expensive, but they are worth it; if you don't buy quality electrical tools, you should buy a good, prepaid insurance policy.

My tools may look worn and torn, but I make sure that they all work well. Whenever a tool becomes worn or

I now use a tool pouch instead of a belt. A freestanding pouch takes the weight off my hip and keeps me from losing the tools every time I bend over or climb into a crawl space.

broken to the point that it can't do its job safely, fix it or replace it. In addition, use a tool only for its intended purpose. Most hinged tools need a drop or two of lubricant about once a month. Some need a lubricant spray to clean out the dust and debris. The lubricant should be nonsticky so that it won't attract dust, but it should leave a slick film to allow for smooth operation.

Tool carrier

When I wrote the first edition of this book, this section was called "Tool Belt." However, I've since changed the way that I carry my tools—now I use a pouch. I'm not sure if I did it to create a more efficient system or because I grew tired of carrying around the extra weight, but the new system works better. With my old belt, it seemed that my tools used to fall out every time I bent over or climbed into a crawl space. It also seemed that I was always taking off the tool belt to do something. The stand-alone pouch has solved all my problems.

My most-used tools: electrician's pliers (Klein), needle-nose pliers, diagonal cutters (called diagonals), and end cutters (staple pullers). Don't start wiring until you have these tools.

The tools in an electrician's tool belt or pouch will vary slightly depending upon his or her specialty. In any case, each tool should have its own place so you can withdraw it without looking. Stock your pouch with only your most-used tools; otherwise, you'll wind up looking like Rambo ready to go to war. I keep less-used tools in a secondary bucket or in my newest play-tool: a little red wagon.

The wagon has been a godsend. I now have a home base for all my tools. Rather than put the tool down on a counter or on the floor, I put it in the wagon, eliminating the time I used to waste hunting for "lost" tools. When I move to a different location, I just pull the wagon and all my tools come along for the ride.

Side cutters

Side cutters, also called lineman's or electrician's pliers, are to an electrician what

a side arm was to a gunslinger of the Old West. Although used mostly for cutting and pulling wire, side cutters can be used for other jobs, too. The blunt, wide nose can twist wires together before a wire nut or any other mechanical splicing device is applied.

Good side cutters are among the most misused of the lineman's tools. Being heavy-duty, they are occasionally used for a hammer and other things I am ashamed to mention. The only way I stopped abusing my side cutters was by keeping a hammer and other miscellaneous tools by my side (normally in the bucket mentioned earlier).

Long-nose pliers

Next to my side cutters, long-nose pliers are my most-used tool. They are great for bending wires into a loop for insertion around receptacle and switch screws and for pulling wires into narrow locations. Never use them to twist nuts off screws; the twisting action will spring the pliers' jaws off to one side and eventually ruin them. Try to find long-nose pliers with an integral wire cutter so that you won't have to change pliers just to cut the wire ends.

Diagonal cutters

Sometimes wire needs to be cut extremely close to the surface of an object so that there is minimal wire protrusion. The thin head of a diagonal cutter allows it to fit into narrow places where other pliers can't go. I often use mine when I have to cut one wire in a bird's nest of wires.

End-cutting pliers

Also called "nippers," end-cutting pliers are used primarily for pulling staples. But they also make good straight-in close-cutting pliers. Sometimes a wire needs to be cut, but access to it is only straight in from the front. Side cutters and needle-nose pliers all cut from the side of the jaw, meaning that the entire head of the pliers must be at a right angle to the wire. End-cutting pliers cut dead ahead.

Wire strippers

There are many different types of wire strippers on the market. If you don't have one, get one. Indeed, I wouldn't allow you on the job site without it. Any wire stripper is good as long as it strips the insulation off the wire and doesn't nick or cut the metal of the wire. Inexpensive manual cut-and-pull strippers are the most common. Automatic wire strippers strip faster, but they are more expensive. Most automatics don't work well on the new NMB cable because of the conductor's nylon sheath. And no automatic wire stripper is designed to fit into tight locations. For such situations, I use a common cut-and-pull stripper. Whichever stripper you buy, get one that strips the most common wires: 16, 14, 12, and 10 gauge. The 16-gauge wires are used in light fixtures; the others are used in common house wiring.

Wire strippers. The yellow-handled cut-and-stripper is one of the most common. The red-handled automatic strippers are my personal favorite. The blue-handled stripper is high quality, but it doesn't consistently strip the insulation off NMB conductors because the cutter blades tend to slip off the nylon sheathing.

Screwdrivers

An electrician needs many kinds of screwdrivers, and they all can't be kept in a tool belt. I keep one high-quality multi-purpose screwdriver (a driver with interchangeable heads) in my pouch and the rest in the tool bucket.

WORK SAFE
WORK SMART
THINKING AHEAD

Never use a knife to remove insulation from solid wire. The knife blade notches—and weakens—the wire, which may cause it to break in the future. For the same reason, you shouldn't use bladed pliers to cut and pull insulation off a solid wire.

Specialty screwdrivers, such as these speed drivers, are nice to have around because they have handles that spin. As you twirl the handle, the centrifugal force of the L-shaped shaft drives the screw. Speed drivers only work on loose screws; hard-turning screws still need a common rigid-shaft screwdriver.

There are many types of screwdrivers: flat head, Phillips head, Torx®, square drive, and offset are the most common ones used by electricians. Offsets are used in tight places where a standard screwdriver is too long. I use mine to remove screws from an appliance when there's only about 2 in. between the screw head and the wall.

Don't use the wrong screwdriver for the wrong job. For instance, never use a straight blade in lieu of a Phillips head. And remember that even straight blades have different blade thicknesses—you'll need both ¼ in. and ⁵⁄₁₆ in.

Eventually, you may want to invest in a few specialty screwdrivers. One screwdriver, available in both flat blade and Phillips head, has a clip at the bottom that holds the screw on the shaft. When you have to extend the screw into an area where your hand can't hold it in place, this driver is worth its weight in gold. A tiny flat-blade screwdriver is handy for releasing wires in push-in receptacles. And don't forget the thread-starter driver—I couldn't do my job without one.

Utility knives

Most utility knives use the same cheap retraction mechanism that jams when it gets full of dust and dirt. However, the new low-cost breakaway blade utility knives work great. I prefer the utility knife to other types of knives because the blade is sharp and thin, and dull blades can be replaced in a snap.

You may prefer to use a standard knife instead. But on the whole, a general-purpose pocketknife makes a poor substitute for a utility knife. The point of a pocketknife is rarely razor sharp.

Electrical tape

The electrician will need both black electrical tape and several different colored

electrical tapes. White tape is used to identify any wire that is a neutral (grounded conductor) and doesn't already have a white coloring on its insulation. Gray may be used for neutral marking, but I have never seen it done. Green tape is used to identify grounding wires, the wires that ultimately connect to earth ground via a ground rod. Any color tape other than white (gray) or green can be used to identify different hot wires if you want to keep track of them. Colored tape should be used for color-coding only. Once you've used colored tape, you will realize that it is not the same quality as good black electrical tape. Black tape is used for general taping purposes—splicing, insulating a metal box, holding things together—as well as identifying a hot wire from the panel or utility.

▌ **ABOVE CODE**　Do not use the cheap electrical tape found in grocery stores and discount houses. I prefer a high-quality tape, such as Scotch® Super 33+ (7 mil thick) or Super 88 (8.5 mil thick). Both are good tapes that work from 0°F to 260°F and resist ultraviolet rays.

For splicing in extremely hot applications above the temperature range of standard tape, I recommend using 3M™ 69 Glass Cloth electrical tape. For covering splices with a solid rubber or silicone rubber tape, use 3M 130C or Scotch Tape 70. All tapes are available through professional electrical suppliers.

Extension cords

Extension cords are valuable tools for anyone working in the trades. Cords come round or flat; I prefer the round ones because they are much easier to loop for storage. I really like the ones with the lighted ends, because you never have to guess whether or not the cord has power.

Typical extension-cord gauges (from thinnest to heaviest) are 18, 16, 14, 12, and 10. A cord with a smaller gauge number, say 10, provides more power than a cord with a large number, say 14.

ABOVE CODE   With heavy-duty tools, I never use any cord less than 12 gauge; however, lighter-duty tools could use 14 gauge. The longer the cord, the heavier the gauge you need. I have two 50-ft. 12-gauge cords, one 50-ft. 10-gauge cord, and one 100-ft. 10-gauge cord. To keep the longest from getting tangled, I wind it around a 10-ft. 2×6 with Vs notched on both ends.

Make sure the cord you buy has enough outlets. You should buy either a cord with multiple outlets or a separate 1-ft.-to-3-ft. cord with multiple outlets. Never use the cheap multiple-tap fixtures that plug into an outlet—they can't take the current flow.

Ladders

Ladders, like extension cords, are necessary for most trades. I have a 4 ft., a 6 ft., an 8 ft., and a long extension ladder—all fiberglass. I chose fiberglass because it doesn't absorb water, it doesn't warp, it is strong, corrosion resistant, and—this is very important—nonconductive. Electricians should never use aluminum ladders unless they want to have a short life. Wood is fine, but it cannot take abuse and weathering the way fiberglass can. My wooden ladders generally lasted less than five years.

Ladders come in different grades, which may be very confusing. All you have to remember is to get a type 1A fiberglass ladder of whatever length you want. (The 1A designation means that the ladder is extra-heavy-duty.) Sometimes you can go by color to determine

Stripping Cable Safely

Stripping the sheath from a cable is an art. You feel just like a surgeon cutting the skin but nothing underneath. To do this operation, keep the cable flat, not twisted, and support it on a solid surface. Never, under any circumstances, support the cable with your knee, as I have done, while cutting the sheath. As soon as you think everything is going fine, the blade will be buried deep in your leg (I have a scar to prove it). And don't try to cut into the sheath while holding it in mid-air. Once the cable is properly supported on a table or bench, use a sharp utility knife and gently slice down the center of the cable. The cable center contains only the bare ground wire, so there will be no harm if the blade occasionally goes in too far. Once cut, peel back the sheath and slice it away.

Slice down the center of the NM cable to strip off the jacket. If you cut too deep, you'll only scratch into the bare ground wire, not cut through the sheathing of insulated wire.

An alternative stripping method. I prefer cutting into the jacket, then tilting the blade to one side. The blade slides over the top of the insulated conductors as it cuts the jacket.

Ladder Safety

Ladders may appear tame, but they are one of the most dangerous tools in this chapter. My left ankle is all steel from a ladder mishap. One minute everything was fine. A second later, I hit the ground and heard the sound of breaking bones. If you value your life or property, take the following precautions:

■ Use a ladder with swivel feet, which prevent the ladder from kicking out when you are working on it. However, the feet won't work if they're just sitting on the ground (left). Stomp on the first rung to sink the claw into the ground (right).

■ When working over a wood deck, clamp or nail a board behind the ladder's feet to keep them from kicking out.

the ladder grade. Louisville ladders, for example, are color-coded: orange for class 1A and yellow for class 1.

There are all kinds of accessories available for ladders, and you will eventually buy a few of them as the need arises. For instance, because ladders leave scratch marks where they touch a wall (indoors or out), you may want to cap the end. A low-cost method is to wrap a towel around the ladder ends that make contact with siding or walls. Another option is to buy special end caps.

AC-Powered Tools

Many types of AC-powered tools are required on a job site. The extra torque provided by drills, saws, and other corded tools—as opposed to cordless, battery-operated tools—is a nice thing to have. Here, I will tell you my experiences with such tools, both good and bad.

Safety is paramount at home or on the job site, and tool maintenance is important to your safety. Follow the manufacturer's instructions for tool use and maintenance, and pay attention to the power cord on the tool. The cords on AC-powered tools get worn and damaged (I've lost count of the number of cords I've seen with damaged sheaths or bare wire showing). These should be fixed or replaced as soon as you notice any signs of wear. And although I shouldn't have to say this, never break the ground lug off a three-wire cord to fit a two-wire outlet—this is compromising safety for convenience. When the time comes to replace the cord's plug, install a high-quality plug as opposed to a dime-store plug. Never use a power tool around water—or even excessive moisture—and always plug into an outlet protected with a GFCI. Finally, never use a power tool

My right-angle drill excels at hole cutting and deep drilling. Its right-angle design, along with a short auger bit, allows it to fit between studs. Its tremendous torque makes it extremely dangerous to those who don't know how to use it.

Drills

Roughing-in wiring in a residence requires heavy-duty drilling. Although most drilling can be done with a ⅜-in. drill and a sharp spade bit, a professional has a wide spectrum of drills and bits. I use two AC-powered drills: a ½-in. pistol grip and a heavy-duty ½-in. right-angle drill. A ⅜-in. drill just doesn't have enough power. Everyday, heavy-duty use would eventually ruin it.

Heavy-duty drilling rigs, however, are very dangerous and should be used only by professionals—and with extreme care. They have so much torque that, without a clutch, they can rip off an arm or break a wrist. I currently use a Black & Decker® TimberWolf, which has a clutch, on low speed, that engages when the bit sticks. It has two speeds: 300 rpm for drilling large-diameter holes and 1,200 rpm for speed drilling. I use the low speed whenever I am using a self-feed or a hole-cutting bit and sometimes when I'm using an auger bit. This drill is what you want all drills to be: heavy duty and well designed. Right-angle drills also come in medium duty, which can also cause injuries. I don't recommend them for the professional because they will invariably get used in heavy-duty situations. My pistol-grip drill is a Makita® ½-in., but don't be deceived by its looks. Being ½ in. and without a clutch, it can break your wrist in a heartbeat just as a right-angle drill could.

Drill Safety Checklist

To operate a right-angle drill safely, follow these precautions:

- Maintain a good, tight grip on the tool.
- Always assume that the drill may twist out and fly around. So keep your head away from any place where it could get smashed by a runaway drill, and don't hold your arms in an awkward position. Also, do not stand in a precarious position to drill. Anchor yourself in a good, solid position.
- Wear soft, flexible leather gloves and safety glasses.
- Feel what is happening through the vibration of the drill. Many times you can pull back as you feel the drill start to work harder, become jammed up, or even hit a nail.
- Never drill with a damaged or dull bit.
- If you feel the bit jam, let go of the trigger immediately. Try not to let go of the drill. If the drill twists out of your hand, jump out of the way immediately, which leads me to the next rule.
- Always have a clear area behind you.
- Because of the last two precautions, heavy-duty drilling on a ladder should be avoided if at all possible.

Spade bits with protruding cutters allow for fast cutting.

Drill bits you'll need

For cutting through wood, I use several types of bits. Which bit you use will depend on the situation and whether you'll need to cut through a double top plate, a corner with nails, or a stud. It is important that you know what each bit can and cannot do so that you can choose the right bit for the right job.

SPADE BITS   There are several spade designs on the market. I've found that spade bits with two protruding end cutters cut faster than the standard flat-bladed variety (see the photo at right). Spade bits come in short (4-in.), standard (6-in.), and long (16-in.) lengths. One problem with spade bits is that they tend to break at the shank if you don't keep the drill centered in the cutting hole. DeWalt solved this problem by making extra-thick shanks. DeWalt also makes a ¼-in. hex shank attachment for hex drivers, which allows the blade to be installed without opening the chuck. Spade bits normally come in ¼-in. to 1½-in. diameters.

AUGER BITS   Auger bits are long-bladed spiral bits used to drill long distances through wood or several studs at once. One problem with long drill bits, and especially with auger bits, is that they tend to bind and get stuck in the hole. When this happens, remove the jammed bit. First, unplug the drill and remove the

Auger bits are used for fast and deep hole cutting. Extension arms attach to the bit via two setscrews, allowing you to cut deeper holes.

drill from the bit. Next, use a small pipe wrench to turn the drill-bit shaft counterclockwise and unscrew the bit. To prevent this from happening in the first place, coat the drill bit's flat sides with wax, soap, or wire-pulling compound. The slick coating keeps the bit from binding on the sides of the hole. Another problem with auger bits is cutting into nails. You'll normally feel a repeating "thunk" as the edge of the cutter keeps cutting into the steel. This won't be a problem if you're using a bit that can cut through steel, such as the Greenlee® Nail Eater II®. If you hit a nail with this bit, the auger will cut through it, instead of jamming the bit, which could hurt the person holding the drill. (Before drilling, try to remove any nails that will be in the way or find another route.)

Auger bits are available in different diameters (¼ in. to 1½ in.) and lengths (6 in., 7⅝ in. or 7¾ in., and 18 in.). I recommend both a long bit (18 in.) and a short bit (6 in.), as well as an extension (see the photo on the facing page). I use 1-in., ¾-in., and ½-in. bits.

SELF-FEED WOOD BORERS The large boring bits that cut away all the wood within the perimeter of the cutting circle are called self-feed wood borers. These bits come either in one solid piece or with blades. I prefer the blades, because they cut very fast and accurately. However, if either one cuts into a nail, look out! The bit simply stops turning. And when this happens, the drill body itself will start to turn, hitting the hapless worker in the head or breaking his or her arm. This type of bit should be used only by professionals—and preferably in a drill with a clutch.

HOLE SAWS The bits that cut only the outside perimeter of the cutting hole are called hole cutters, or hole saws. I started using hole saws in lieu of self-feed bits because I had so many close calls. I use only carbide-tipped hole saws (special order), but the most common type is bimetal. I prefer the carbide tooth because it can cut through a nail as if the nail weren't even there. A nail will ruin a bimetal cutter.

Self-feed wood borers cut large-diameter holes quickly. The bit on the right has replaceable cutting blades. Such bits should only be used in a heavy-duty right-angle drill.

Hole saws are safer to use than self-feed bits. They come with and without an arbor.

Hole saws come with and without an arbor (the part that holds the drill bit). The arbor fastens to the cutter head, and then fits into the drill chuck to transfer the drill's rotating power to the cutter head. I prefer to buy a hole saw without an arbor and then buy the arbor separately. That way, I don't have to throw the arbor away every time I buy a new hole saw. Hole saws that come with an arbor are more expensive than those without an arbor are. One arbor can be used with a number of hole saws, which keeps the cost of the cutter heads to a minimum. Carbide-tipped hole saws are available in sizes ranging from ¾ in. to 6 in.

An option to the two-part cutter-head and arbor system is the one-piece arbored hole saw. These hole saws tend to cost more because you have to buy the arbor each time you buy the hole saw. I don't use them much because they are normally only bimetal tipped.

Either system, whether with an integral arbor or with a separate arbor, will require a pilot bit, which centers the hole saw in the hole to be cut. Pilot bits have a habit of breaking, so have several spares on hand. Lennox recommends using a carbide-tipped pilot bit; I don't. A standard twist-drill bit cuts into the wood faster and is cheaper to replace if the bit breaks.

Twist drills are commonly used for drilling wood and steel.

TWIST DRILLS Standard steel-drilling bits (called twist drills) are needed for drilling wood and steel. A common application is for drilling pilot holes through a floor and into the crawl space to act as a position locator. For this job, a 12-inch-long ⅛-in. to ¼-in. bit works fine. I've found I don't need the complete set of bits down to $^{31}/_{32}$ in. that are sold in hardware stores. I normally use ¹⁄₁₆ in., ⅛ in., ¼ in., ⅜ in., ½ in., and ¾ in.

Saws

I use corded saws on the job for heavy-duty cutting. The circular saw does the common work of cutting straight edges when I'm trying to cut through plywood or heavy wood. My primary AC tool for cutting anything is the old standby, the reciprocating saw, commonly called a

Sawzall®, named after the original design by Milwaukee.

Currently, I use Milwaukee's low-vibration Super Sawzall and its cordless version. I've been using these tools for several years, under hard work and neglect, with no problems. I cut wood, nails, steel, cast-iron pipe, rebar—anything I want.

It is with the reciprocating saw that you'll make most of your accidental cuts through hot wires. That's why it's very important to have a double-insulated tool or, as in the Milwaukee, a grounded tool. In addition, the Milwaukee has a rubber boot over the front half that you hold when pushing the blade into the wood. I've cut many hot wires without even a minor shock. I've lost only the blade—and the wire.

Besides the standard blades for reciprocating saws, extra-thick and bimetal types are also available in varying lengths. The reciprocating saw makes only rough cuts, because the blade tends to move around in the cut. Some blades move more than others do; extra-thick blades seem to track better. I normally use the bimetal type because it has some bend in it; it gives a bit without immedi-

ately breaking. That blade costs a little more, but its longer life more than repays the investment.

A good jigsaw is another important tool in the electrician's arsenal. This is the tool for straight cuts, as compared to the reciprocating saw and its wandering blade. I don't like those saws that use a hole-in-the-blade design. They seem to break right where the hole is. I prefer tools with a bayonet design—they break, too, but not as easily. You'll need a variety of blades, but you'll break fewer of them, and be safer to boot, if you let the jigsaw cut at its own speed, with only slight pressure from you.

Cordless Tools

My first experience with battery-powered tools was unpleasant. First-generation cordless tools were poorly designed and useful only for very light-duty applications. My distaste lasted for many years. Today, the market is full of high-quality cordless tools. Now I'm not without my battery-operated drill, saw, and screwdriver. Cordless tools offer a few advan-

> ▪ **WORK SAFE**
> ▪ **WORK SMART**
> ▪ **THINKING AHEAD**
>
> Never cut with any saw, especially ones with long blades, unless you know what is behind the wall. It could be plumbing lines or electric cable—even the main service-entrance cable. If you cut into the latter, you and the saw will be in big trouble, because there may not be an overcurrent device (breaker) in the system, except what the utility has on the pole.

My three corded saws are the reciprocating saw, circular saw, and jigsaw. I've put all three of these through extremely heavy-duty use, and they are still working fine.

I use myriad cordless tools. From left to right: Makita mini circular saw, Milwaukee screwdriver, Hitachi medium-duty drill, DeWalt circular saw, Makita mini reciprocating saw, and Milwaukee reciprocating saw.

tages specific to each one. The most obvious, of course, is that the tools don't require an extension cord, so they are very portable and won't electrocute you.

All cordless tools need battery chargers. In the past, the battery needed about an hour to charge. Today, some chargers require only 15 minutes to do the job. For continuous work, you'll need several batteries on hand. I'd stay away from the charger design that requires you to push a button to start the charging cycle. You may come back later to find that the battery isn't charged because you forgot to push the button or that you pushed it, and it didn't work. A good charger indicates when it's charging and when it's done. It also analyzes the battery to see if it has any problems. For example, it will tell you if the battery is weak and cannot supply full power.

Drills

Cordless drills, when used with a sharp bit, can drill most of the holes in the house and won't break your arm doing it, as a heavy-duty AC model may. And they can't electrocute you, even if they cut into a hot wire, because the case is made of plastic. I do at least 90% of my drilling

with cordless drills. They offer me speed, versatility, and safety. I can use a cordless drill on a ladder, around water, and in crawl spaces and other awkward places where I would normally have to worry about pulling, dragging, or lifting an extension cord.

Cordless drills are available in several different voltages ranging from 6.9 volts to 14.4 volts and higher. The 9.6- and 12-volt units are the current favorite for the average do-it-yourselfer because they are lightweight and have a decent amount of power. The more serious do-it-yourselfer, or the professional, may need a higher-voltage, larger-chuck unit.

Today's cordless drills are available with ⅜-in. and ½-in. chucks. I use the ½-in. one because it can accept larger-diameter bits. Most cordless drills now come with keyless chucks. At first I couldn't wait to get them, and now I wish I hadn't. Keyless chucks are great for changing bits quickly, but beware of their minor problems. Without using a chuck key, you can get the bit only so tight, and if you're drilling deep, the bit may stick in the wood and start spinning in the chuck. In addition, to tighten the bit in the drill, you have to place your hand on the drill chuck head and pull the trigger. The chuck tightens until the motor stalls. Once the motor stalls, you have to assume the bit is tight in the chuck. This process wastes valuable battery power.

Circular saw

Another battery-operated tool that I use often is my DeWalt cordless circular saw. This is not a toy—it cuts with the best of them. It works like a champ to open up areas and to make fast trims and cuts. Carpenters love this saw for overhead cutting. It's nice just to pick up the saw and cut rather than look for a grounded outlet and drag out the extension cord. If you're working in areas where no electricity is available, this is a good saw to have around.

Screwdriver

In my opinion, the Milwaukee 6538-1 cordless screwdriver is the best on the market. It's heavy duty, unlike the other play-toys sold at hardware and cut-rate stores. I use it primarily for getting into multigang switch and receptacle boxes. A four-gang switch box has 16 screws to pull before you can get into the box.

Specialized Tools

The tools in this section are very specialized. I don't recommend that people doing minor wiring jobs buy them, because many of them are expensive and you may only use them once or twice. Instead, these tools are for professional electricians and serious do-it-yourselfers (those folks who plan on doing some big-time wiring work around the house).

Hole punchers and stepped drill bits

All electricians know that the prepunched holes in meter bases and control panels are rarely where you want them. A lot of labor and material can be saved if you place the conduit and the cables where

My two favorite cordless drills. I use the Hitachi for common everyday drilling and when I'm on a ladder. I use the Milwaukee for heavy-duty drilling.

you need them, as opposed to where some desk jockey decided you needed them. There are many different ways to punch holes in metal boxes. The way I have settled into is with a Cobra 400:1 gear-reduction unit that uses Greenlee Slug-Buster® hole punchers. First, drill a pilot hole through the steel. Then fit the gear-reduction shaft through the hole with the Slug-Buster on either side. Attach a drill to the unit's shaft and pull the trigger. You can even use a cordless drill for this operation. The rotating shaft pulls

the Slug-Buster together and punches the hole faster than I can explain it. I use a standard hole-punching kit that goes from ½ in. to 1¼ in. I also purchased 1½-in. and 2-in. Slug-Busters. As opposed to punching holes, special hole-cutter-type drill bits are available to cut straight through the metal.

For drilling small-diameter holes quickly, such as the pilot hole for a hole puncher or a knockout, nothing beats a stepped drill bit. This is a special metal-cutting bit that cuts various diameter holes depending on how far you push the bit into the metal (the farther you drill, the bigger the hole). Stepped drill bits are available in two designs: a blunt-end variety used to increase the diameter of existing holes and a tapered-point variety used to open new holes. I use either one to increase the diameter of the knockout holes (up to 1⅜ in.) in metal boxes. I also keep carbide-tipped hole cutters around for very thick steel.

Conduit benders

Professionals bend conduit, while do-it-yourselfers use premade angle fittings. I keep ½-in. and ¾-in. benders on the truck for just those times when I'm allowed to have fun . . . I mean, bend conduit. However, mastering the art of bending conduit takes a tremendous amount of time, patience, and money (for wasted conduit to practice with). If you want to install the conduit fast, forego the conduit benders and use preformed angles.

Fish tapes

Pulling and fishing wire can be a major problem if the wrong equipment is used. The purpose of a fish tape is to "fish" wire through walls. Here's how: Put the thin metal tape into a hole in a wall, maneuver it through the wall, and bring it out a different hole. Then attach a wire

The Greenlee hole puncher kit. Consider buying 1½-in. and 2-in. hole punchers in addition to this basic kit. To use the puncher, you'll need to drill a starter hole for the bolt.

Nothing beats a stepped drill for drilling holes through steel. It's commonly used to drill holes in service panels for NM connectors.

to the end of the fish tape and pull the wire into the wall as you pull the fish tape out.

However, a standard metal-coiled fish tape is, in my opinion, the wrong tool to use. It is conductive, has an end that catches on everything in the wall so you can't get it back out (many times I have had to cut the metal fish tape and abandon it in the wall because it had caught on something), and is hard to get through a wall or ceiling cavity because it always tries to go in a circle. To counter this, I cut several feet of metal fish tape off my reel, bend it straight (I store it straight as well), and cut off the looped end. Now it easily slides inside wall cavities and through insulation. I don't recommend using fish tape made of metal, and the nylon tapes I've seen stay circular just like the metal ones do, and they catch stuff in the wall cavity. However, I hear the fiberglass tapes work well.

What makes a good fish tape? It should be nonconductive and flexible but just rigid enough to stay straight in the wall. A 20-ft. straight length of ½-in. or ¾-in. polybutylene plumbing pipe or a semiflexible equivalent, such as PEX pipe, makes a perfect fish tape. It is rigid enough to go long distances within walls and ceilings and stays flexible enough to bend into the cutout hole and pull through. Once through, you can push the wire up into the pipe and tape it off. As you pull the pipe back, out comes the wire. This is the method I use most often. I normally use ½ in. for the walls and ¾ in. for the extra-long runs in the ceiling. To prevent the end of the pipe from catching, I tape or insert a marble at the end.

Rotary hammer

Used to break up and drill concrete and punch through cinder block and brick walls, the rotary hammer (also called a

A rotary hammer is the tool to use for getting through concrete walls and driving ground rods. This type of hammer drill uses spline bits (see the photo at right). Each bit costs around $20 to $100.

hammer drill) is an indispensable tool for the professional. Whatever brand you select, make sure you get one with a clutch that engages when the bit sticks, because this type of drill has tremendous torque. I also use the drill for installing ground rods. Makita, Bosch, and several other manufacturers make a special attachment for their hammer drills that will drive a ground rod into the soil in just a few seconds. It sure beats using a sledgehammer.

VOMs and multimeters

Volt-ohm-meters are the most valuable tools in any electrician's tool chest. I still call them VOMs because that's what we used to call them in the early days. Technically, in this day and age, they're called multimeters. Having been through many of these devices—from the old analog Simpson 260®s to today's digitals—I have to say that my digital Fluke® model 25 and model 30 are the best I've ever used. They do everything with extreme accuracy, and they are almost indestructible (I've dropped mine on concrete and they still work). Some digitals

Two tools I cannot do without. The yellow meter is autoranging—you don't have to turn a knob to an approximate voltage—and measures both AC and DC. The black meter measures only AC. Its primary purpose, however, is to measure current. The fingers clamp around a wire and the meter gives an indication of the amount of current in that wire.

board heaters to verify how much current they're pulling without getting near a bare wire. The model 30 is also a multimeter that measures voltage, resistance, and continuity (measuring continuity is a way of verifying that two points are electrically connected). However, it is not autoranging, meaning that you'll have to know the approximate voltage of the circuit and switch it to the proper scale. I prefer having two multimeters on hand—one as a backup.

There are many accessories available for VOMs: extra-long test leads, alligator clips that allow one or both hands to be free, and line splitters. Line splitters allow current to be checked in any corded appliance.

Plug-in analyzers

I use plug-in circuit analyzers and GFCI testers to test if a receptacle is wired properly and whether or not a GFCI is still providing ground-fault protection—you'd be surprised to know how many are not. But there are situations when you cannot test a GFCI using a plug-in tester. The bottom line is that whenever possible, always use an integral GFCI test button.

In the past, a GFCI tester was simply a plug-in tester with a push button on it to simulate a ground fault. No longer. State-of-the-art testers can now give you a whole world of circuit analysis from a little plug-in detector. Although I have several of the little cheap ones, the one I use most is the SureTest® Circuit Analyzer (from Ideal Industries, Inc.). This unit is too elaborate and expensive for the amateur but should be in the arsenal of every electrician and official who does a lot of testing.

Besides standard wire polarity checks and GFCI testing, the SureTest also tests for the presence of a bootleg ground. A bootleg ground occurs when someone

make you wait while they count up to get the voltage. However, both of these models give an instant voltage reading. They also work in cold weather (some models don't).

The model 25 is autoranging, which means that I don't have to know the voltage before I measure it and switch it to the right scale. I simply put the probes on the testing points. It has a built-in continuity tester to test lightbulbs and water-heater elements, a capacitance tester to test motor-starting capacitors, and a millivolt meter, which is useful when testing thermocouples on gas water heaters.

The model 30 clamp-on meter indicates how much current the branch circuit is pulling without having to open the circuit. I can check submersible pumps, water heaters, and electric base-has incorrectly installed a jumper wire on the receptacle back from the neutral over to the ground lug (a common tester will give you a false "OK" reading).

I also use this tester to check the resistance along the equipment-grounding conductor. That is, if you plug it into a receptacle, it will give you a resistance reading along the bare ground wire from the receptacle back to the breaker panel. This test is very important if you are doing any lightning suppression. A surge suppresser—for example, the one at a computer—takes the surges off the line and sends them back to the panel and earth via the equipment grounding conductor. If you have a loose screw on a receptacle or a bad splice, the suppresser won't work properly. If the test gives you a high reading, then you know there is a problem.

Plug-in testers indicate if an outlet is wired correctly. The bigger device is more expensive, but it also performs additional tests, making it a worthwhile investment for a professional.

Safety Equipment

The most dangerous problem on the job site—besides electrocution—is debris going into your eyes, which makes approved safety glasses an electrician's best friend. They must be comfortable, or you will be tempted not to wear them. Be sure to get some that have adjustable arms. Antifog and antiscratch coatings are nice, but they fog and scratch anyway. I hang my glasses from the rearview mirror of my truck, which reminds me to put them on when I leave the truck. Get into the habit of hanging your safety glasses around your neck when they're not in use. I've found that if I don't have my glasses within arm's reach when I need them, I won't use them.

After your eyes, your toes are the most in danger. Always wear steel-toed shoes. Falling drywall, plywood, and heavy control-panel lids are always threatening to amputate your toes or injure your feet. If you're around a job site long enough, something will happen. If hard-soled steel-toed shoes are not to your liking, steel-toed running-type shoes are now available.

Rubber gloves, soft leather gloves, a rubber mat, and a hard hat are items that should also be kept with you—or at least on your work truck. Rubber gloves can save your life. Once I was wiring merrily away in a panel—nowhere near a bare hot wire—and zap! What happened? I discovered that I had brushed against a hot wire where a previous electrician had gouged off a section of insulation. From then on, I started wearing rubber dishwashing gloves in hot electrical panels.

CHAPTER 3

The Service Entrance

THE SERVICE ENTRANCE (SE) is the part of the electrical system that starts at the utility transformer and terminates at the main service panel. It includes the cable from the transformer to the residence, the meter base, and the service panel. A residence has either a buried or an aerial service entrance. In either case, part of the cable is provided and installed by the utility; the rest is your responsibility. In a *buried service,* the utility provides the cable and labor to install the service entrance all the way to the meter base. In an *aerial service,* the wiring that swings from the pole to the house is called a service drop, or triplex, and is provided and installed by the utility. The utility's responsibility stops at the point where the drip loop connects to the house's service cable. In both cases, the electrician installs the meter base.

Many initial issues in new residential construction concern the service entrance, which is a major factor in the design and construction of a home. Coordination and cooperation between the utility, the local inspector, the building and electrical

contractors, and the homeowner is mandatory during this time to avoid problems and errors. This chapter will help you make intelligent decisions regarding the design and installation of the service entrance.

Choosing a Buried or an Aerial Service

The first decision is whether you want a buried or an aerial service entrance. (If you don't request a particular type of service, the engineer will figure the cost of both and put in the cheaper option.) If you choose the more expensive option, but it is within a set amount, you pay nothing; however, if it is over that amount, you may be expected to pay the difference. In some cases, you can avoid that added expense by doing some of the work yourself. For example, with buried cable, the utility usually takes care of opening the ditch, but you could offer to open and close the ditch yourself.

Pros and cons of each service are listed on p. 52. Whichever choice you

The Service Entrance

Aerial Service Entrance

Neutral

Point of attachment

Hot primary

Service drop, or triplex

Utility responsibility stops at splice.

SE cable

Ground wire

Meter base

Buried Service Entrance

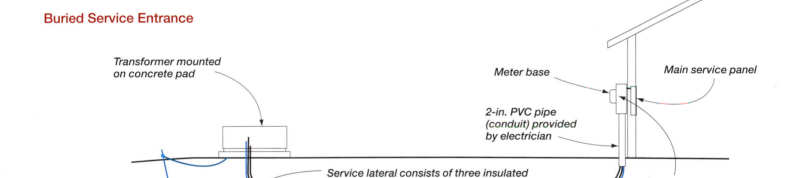

Transformer mounted on concrete pad

Meter base

Main service panel

2-in. PVC pipe (conduit) provided by electrician

Service lateral consists of three insulated conductors buried 4 ft. in ground.

Utility responsibility stops at top lugs inside meter base.

Buried Service Entrance

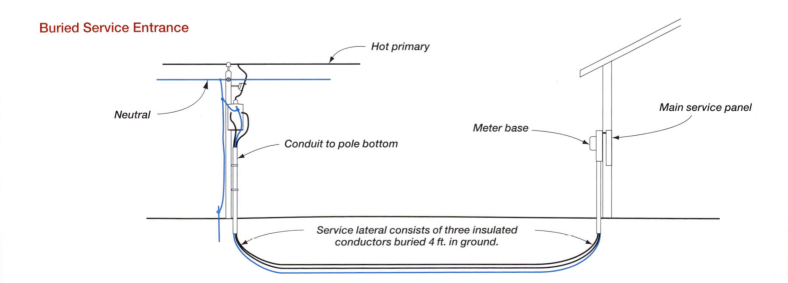

Hot primary

Main service panel

Neutral

Conduit to pole bottom

Meter base

Service lateral consists of three insulated conductors buried 4 ft. in ground.

Buried service entrance requires a deep ditch—expect things to be quite messy. The utility service entrance cable is lying to the right of the ditch.

Buried vs. Aerial Service

Advantages of Buried Service

- Fewer trees to cut
- Smaller right-of-way requirements
- House looks better without exposed conduit and cable and utility poles
- SE protected from ice storms and falling trees
- No clearance problems
- Utility company provides cable to meter base
- No bent masts
- No cables crossing roof
- No damage to roofline (aerial service gets extra tricky if you have to insert a mast through a metal or tile roof)

Advantages of Aerial Service

- Won't damage your landscaping
- Easy to see potential SE problems
- No chance of cutting through buried utility
- Aerial lines can span boulders, creeks, and swamps

make, you need to meet with the utility engineer and discuss where the utility cable, and perhaps the utility poles, will be placed on the property. In order to do that, you must know on which side of the building you want to place the meter base.

Calculating Amperage and Choosing Cable

As soon as you know whether you're going with a buried or an aerial service entrance, your next task is to figure out the amperage size of the residence's service. You can guess the loading simply by adding most of the heavy-duty loads, which is what most people do. However, many times, pros are required to do a load calculation. In addition, the inspector may want to see a load calculation before he or she will issue a permit. The best reason to do a load calculation is to ensure that you don't underestimate your needs and install a service that's too small.

The official electrical load of a residence is sized according to NEC guidelines. Once you know the load of the house, you can choose the correct-size meter base, cable, and main panel for the amperage of the house. In Appendix D of the NEC, you'll find formulas for figuring the load of a residence. Here's how I figure the load for a 2,000-sq.-ft. house:

1. Calculate the square footage of the house, excluding uninhabited areas, such as an unfinished attic, garage, and carport, to obtain the general lighting requirements. Multiply the square footage by 3 to obtain the power required: $2,000 \times 3 = 6,000$ watts, or VA.
2. Add 1,500 VA for each small-appliance circuit in the kitchen. A minimum of two is required

Making Way for the Service Entrance

The route the utility cables take to get to the house should be decided upon before construction begins. In some locales this is no big deal—for instance, when you are adjacent to a utility line. In the country, however, it can be a very big deal. You may be quite a distance from the utility. If the cable crosses another property, a right-of-way will have to be obtained by the utility from the owner. You may also have to sign a right-of-way to get it across your property.

A right-of-way is a legal document that gives the utility permission to install and maintain equipment on your property or on someone else's; normally the utility will not install its lines without the form. It is signed and filed in the county courthouse. Once the form is signed, the utility, not you, controls the section of property specified in the right-of-way (but you still pay the taxes). If the utility has to clear trees and such from the right-of-way, they may charge you for it—therefore, it is common for the landowner to clear it.

The width of the area to be cleared depends on whether primary (high voltage) or secondary (120/240 volt) power lines will be passing through it. An aerial primary line needs at least 40 ft. of clearance. A secondary line normally requires 20 ft. or less, depending on the utility. Whether it will be a primary or a secondary line depends on the length of line and the voltage drop. The utility engineer will give you a report.

A buried service entrance requires only a path wide enough for equipment to pass through. If placed along the driveway, no additional clearing of trees may be needed.

Aerial Lines through Woods

A primary line requires a 40-ft. clearing through woods. A secondary line requires 20 ft.

Buried Service

A clearing through the woods needs to be only as wide as a trencher or backhoe. Other equipment can go around trees or boulders.

by code, so the number will be 3,000 VA. Then add 1,500 VA for the laundry, excluding the dryer. Total = 4,500 VA.

3. List all appliances to be included and write down their power ratings in volt amps or watts. Some common appliances include a dishwasher (1,500 VA), water heater (4,500 watts), clothes dryer (4,000 watts; do not use the standard rating of 5,000 watts for the dryer—use the nameplate rating for the load calculation; the same goes for the stove), stove (8,000 watts), water pump (1,500 VA), garbage disposal (900 VA), whirlpool tub, and so on. Don't include the air conditioner or the heating unit. Let's assume that the total in this case is 30,000 watts or VA.

4. Add the numbers up: 6,000 + 4,000 + 30,000 = 40,000 watts.

   Because all loads don't operate at the same time, take the first 10,000 at face value, then take 40% off the rest: 10,000 + [.4(30,000)] = 10,000 + 12,000 = 22,000.

5. Add the power of only the heating unit or the air conditioner, whichever is the larger, and add it to the total load. In this example, the heating unit is 20,000 VA. Therefore, 20,000 + 22,000 = 42,000.

6. Divide the total wattage by the service voltage to find the total electrical load of the house: 42,000 watts/ 240 volts = 175 amps.

ABOVE CODE This installation requires a 200-amp main panel and meter base. But since we are only 25 amps from maxing out the panel (the voltage of one hair dryer or window-fitted air conditioner), I recommend investing in a larger service to ensure that the homeowner has enough amperage not only for the present load, but also for future loads. In general, I never install anything smaller than a 200-amp service (it is against code to install anything less than 100). Compared to the smaller services, the cost difference is insignificant. For loads greater than 200 amps, round up to the next 100-amp increment, such as 300 or 400 amps.

Choosing SE cable

The SE conductors and meter base must match the ampacity of the house's electrical system. This breaks down into two sections: drip loop to meter base (aerial service only) and meter base to service panel. When dealing with larger services, the big cables that are used from drip loop to meter base are very expensive and very hard to install. This alone should convince you to choose buried. With buried, the utility provides the expensive cable and does the work.

You'll also have to decide whether to use aluminum or copper SE. Copper style

Drip Loop to Meter Base

Size of Service	Size of Copper Conductor	Size of Aluminum Conductor
200 amp	2/0	4/0
300 amp	250 kcmil	350 kcmil
400 amp	400 kcmil	600 kcmil

Meter Base to Service Panel

Size of Service	Size of Copper Conductor	Size of Aluminum Conductor
200 amp*	2/0	4/0
300 amp (200-amp panel and 100-amp panel)	One set of 2/0 and one set of #4, respectively	One set of 4/0 and one set of #2, respectively
400-amp (two 200-amp panels)	Two sets of 2/0	Two sets of 4/0

* Never install a singular panel above 200 amps. A typical 300-amp panel, even a 400-amp panel, has only 40 circuits. A house that needs a 400-amp service panel will need an abundance of circuits—more than what you would want to squeeze into a single panel. You'll need to split the cables into two panels in order to create enough room for all the circuits.

200- to 400-amp Service Entrance

200-amp Service

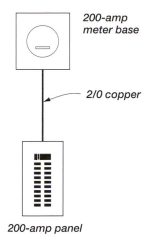

200-amp meter base

2/0 copper

200-amp panel

300-amp Service

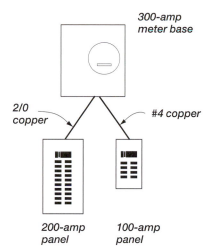

300-amp meter base

2/0 copper

#4 copper

200-amp panel

100-amp panel

400-amp Service

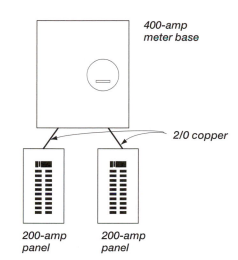

400-amp meter base

2/0 copper

200-amp panel

200-amp panel

U (the flat cable) is out there but may be hard to find (call around to various electrical distributors). I prefer copper service entrance, because it is easier to install two-sizes-smaller gauge for the same amperage and because copper corrodes less than aluminum. Copper THHN/THWN in individual conductors is quite common and available at most electrical supply houses.

Meter bases

Meter bases normally come in 100 amp, 300 amp, and 400 amp. The 400-amp meter base is quite large and heavy and has bolt-on lugs for all the cables—even for the meter itself. The lugs, which are sometimes sold separately, may set you back an extra $5 to $10.

If you have two service panels, you will need to run two sets of SE cables

About the size of two 35mm film cartridges, this two-cable lug bolts onto a 400-amp meter base. The lug attaches onto a bus immediately below the meter itself. You'll need three of these (hot, hot, and neutral) if you are installing two 200-amp service panels.

from the meter base (see the drawing on p. 55). For multiple cables leaving the meter base, you can buy a two-hole lug-bolt kit that accepts two cables side by side—even aluminum and copper. Don't forget to ask for them when you get the meter base.

Local Requirements

Once you've decided where the cable is going to run on the property and what size service entrance you need, the next step is to ask about the requirements of the local utility and inspector. These change from utility to utility and from locality to locality—sometimes even from inspector to inspector. Your area may have a specific system that you must follow to obtain service (an electrical hookup). For example, in my area, the first step is to call the utility and log a request in their system. They'll ask if this is a new installation or an upgrade, the amperage of service entrance for which you want to apply, whether it's buried or aerial, and other questions as well. Some utilities may require a deposit to defray their costs and ensure that you complete the work. The utility will then give you a service request number and issue you a proper-amperage meter base. When that's completed, your next step is to apply for a permit at the local inspection department. (Be sure to call ahead for an appointment.)

Once all the paperwork is done and you've obtained a permit, you can install the service entrance. For a buried entrance, that means from the meter base to the service panel. For an aerial entrance, that means from the drip loop, through the meter base, and to the service entrance panel. To install the service entrance, you will have to "borrow" elec-

tricity from a neighbor, use a generator, pay for a temporary power hookup from the utility ($100 to $500), or use battery-operated tools.

You are normally allowed to install one GFCI-protected 120-volt powered receptacle in addition to the service panel. This receptacle is usually installed immediately adjacent to the service panel. Once all this is done, call your local inspector for a service entrance inspection. The inspector will verify that everything is done within code. If you pass, he or she will notify the utility and they will put you on a waiting list. From that point on, it can take several days—sometimes months—before you get your utility hookup.

Once you have utility power, you are allowed to use that one GFCI receptacle for general-purpose tools. Do not heat up any other 120-volt receptacles until the final inspection. When you have run all the wire within the walls and in the outlets, you are ready for a rough inspection. When the finished walls are up and the receptacles and switches are installed and powered, you are ready for a final inspection.

Locating the Meter Base

Before the utility runs the service drop to the house, you must install the meter base. If the meter base has to be installed before the finished siding is up, attach the meter base to the rough wall and allow the finished siding to be cut around it. Another option is to shim out to the exact distance of the proposed siding and install the siding behind it later. If brick siding will be installed on the house, the meter base will have to wait until the brick is installed. Fasten the box into the mortar between individual bricks. Any type of concrete fastener will work.

Regardless of whether you choose a buried or an aerial service entrance, always try to locate the utility meter base and the main service panel back to back. This design is less expensive and will be less of a problem when it comes time to run cable to the service panel. If the panel is installed several feet away from the meter base, a cutoff panel (which is expensive and will require additional work to install) may have to be installed next to the meter base. The cutoff panel is simply a main disconnect switch, or circuit breaker, that provides fault protection for the SE cable on its way to the main service panel. Without a cutoff panel, if the cable is damaged—say, by a carpenter driving a nail into it—before it reaches the breaker, the only way to shut off the power is at the utility cutoff at the power pole. By the time the utility cutoff removes the power, all the service cable will probably be damaged beyond repair, and it may even start a fire. When such a fault occurs, the wires hum loudly, the utility meter power-usage disk whirls so fast you can't see it, and the utility transformer buzzes. (This is what I've heard; it hasn't happened to me . . . yet.)

Mounting the meter base

The meter base is either mounted at the exact height required by the utility (around 5 ft. from final grade), or in absence of a spec, at head height for easy reading. Before you begin work, double check to make sure you have the right meter base—buried for buried, aerial for aerial—and the right amperage for the job. In the meter base, you'll see four ¼-in. knockouts for four screws. Use a punch to remove the knockouts and be sure to remove all of the metal (check behind the box, because it may bend around back). Remove a knockout from

Back-to-Back Meter Base and Panel

Exterior siding

Interior wall

Main panel

Meter base screwed into exterior siding

When the meter base and service panel are installed back to back, a short length of threaded conduit, either galvanized steel or PVC, can be used to pass wires from the base to the panel.

2-in. hole in meter base

2-in. hole in panel

Threaded nipple

Plastic bushing screwed onto threads

Grounding locknut required for metal nipple

Cut a 3-in. hole through exterior siding for nipple.

Punching a New Hole into a Meter Base or Panel

1

If the meter base knockout doesn't align with the service panel knockout for a nice straight conduit run, you'll need to punch a new hole (normally in the gutter of the service panel). Here's how to do it:

Use a stepped bit to drill a ½-in. hole in the box where you want to position the knockout. Make sure you allow extra room around the hole for the locknut to turn.

2

On each side of the new hole are the two primary parts of a punch: a cutting head (right) and a receiving head (left). Not shown is a bolt that comes with the knockout when you purchase the set.

3

Insert the bolt through the receiving head on the opposite side of the panel and the bolt end through the panel hole. The hollow side of the receiving head should face the panel.

4

Screw the cutterhead onto the bolt until it's snug against the metal, then tighten the bolt with a socket wrench. As the bolt is tightened, the cutterhead will slice a hole through the panel. This 2-in.-diameter hole is for a 2-in. conduit.

the back of the meter base for the conduit that will be the pass-through for the cables running to the main service panel. (For the purposes of this discussion, assume the main panel is located back-to-back with the meter base.)

If you plan to exit the back of the meter base to get to the main panel, you'll want to knock or cut a hole in the exterior wall before installing it. To do this, place the panel top on the 5-ft. mark, use a magnetic torpedo level to make sure the panel is level, and then trace the knockout on the wall (see the photo below). The radius of this circle must be increased by ½ in. to allow for the conduit hub. (If the proposed hole cuts into a stud, shift the base slightly

to go around it.) To cut through wood, use a drill and/or a reciprocating saw. For concrete, I use my Makita hammer drill.

Once the exterior wall is opened, install any adapters on the panel that may be required (see the photos on p. 60). For 200-amp service, the diameter of the conduit and fittings is normally 2 in.; larger service will need a 3-in. or 4-in.-diameter conduit. I prefer to have the buried utility cable enter on the left side of the meter-base bottom instead of in the center. This keeps the utility cable on the left side and leaves ample room to run the exiting SE cable out the right side.

Mount the meter base against the exterior wall with four screws or bolts (do

WORK SAFE
WORK SMART
THINKING AHEAD

Before you mount the meter base, think about where you want the service panel to be, so that the two can be installed back to back. If the SE cable knockouts on the two panels don't align, you won't be able to connect them with a straight piece of conduit.

With the panel top placed 5 ft. above finished grade, place the meter base against the siding and draw a circle around the knockout hole. Once opened, this will be the entry into the house.

Mounting a Meter to a Block Wall

Service panel back against 2× spacer

Meter base interior

Male adapter

Locking ring inside

2-in. PVC conduit

Bushing over threads

Service panel interior

Treated 2× spacer nailed to block wall

Where required, insert a male adapter into the meter base from the back so its threads extend fully into the box. Once in, screw on a locknut and tap-tighten it with screwdriver and hammer.

Screw a bushing onto the male adapter's threads. This bushing protects the cable from the thread's sharp edges.

not use nails—they can work loose). I use 5/16-in. hex-drive screws. Once the meter panel is secured, put a bead of clear silicone caulk along the top and sides and between the panel and the wall to prevent rainwater from running behind the panel and into the cut hole.

Buried Service Entrance

A buried service is the easiest, and sometimes the cheapest, to have installed. To obtain buried service, all you have to do is request it. The utility will normally send out an engineer to make a drawing and figure out what the price (if any) will be. Before buried service goes in, you will need to locate the water line to the well, the sewer line, the septic tank, the distribution field, and the buried telephone line so they don't get damaged.

For buried service, the utility buries from the closest aerial pole or installs a transformer close to the house. The transformer is usually inside a large green box mounted on a concrete pad (see the middle drawing on p. 51). The transformer serves the same purpose as the transformer mounted on the utility pole—it changes the higher transmission voltage to a lower voltage for a residence. The utility is responsible for getting the wires, which are called the service lateral, from the transformer to the actual connection at the meter base. The homeowner, the contractor, and the utility company must agree upon the location of the transformer.

The meter base and the connecting SE cable should be in place before the utility arrives to install the buried service. The electrician must provide the utility crew with a section of PVC conduit long enough to extend from the bottom of the meter base at least 2 ft. underground (approximately 5½ ft. total), a male adapter, two locknuts, and sometimes a long sweep at the bottom. The conduit attaches to the bottom of the meter base with a male hub, two locknuts, and depending on the inspector, a bushing to provide a more rounded surface for the cable to bend against.

Once you cut the conduit to length, simply lean it up against the meter base

Make sure you leave a piece of conduit (plus male adapter, two locknuts, and a bushing) for the utility contractor to install when he or she brings in the buried SE cable. Simply lean it against the meter.

and wait for the utility company to finish the installation. Do not glue the male adapter onto the long sweep, because the utility crew may need to cut the pipe to length as they install it. The conduit diameter is normally 2 in. for a 200-amp service and 3 in. to 4 in. for larger services.

Aerial Service Entrance

Aerial service can take an entire day to install even when you know what you are doing. You must take into consideration the height of the utility cable, porches, and decks, as well as whether or not you're crossing the existing roof. Of course, there's the added safety factor—you will be doing most of the work on a ladder or on the roof.

There are several ways to interface the house SE cable with the utility cable at the drip loop, and your area may require a specific way that is not required by code. Regardless of the interface, your first concern is establishing proper clearances.

Aerial clearances

In Charlottesville, Virginia, a girl was killed when a low-hanging wire caught her by the neck and knocked her off a hay wagon. This tragedy might not have happened if the wire had been at its proper height. After that tragic death, I was hired by the utility to check for other low clearances. I found a bookful.

What people don't know (and the NEC doesn't emphasize) is the difference between being *built* to the proper clearance and *maintaining* a proper clearance. I'm sorry to say that even I learned this lesson the hard way. Once, when installing some cable over a city street in Mathews, North Carolina, I had my two linemen build the line as I checked the pavement-to-cable sag. I measured exactly 18 ft. at the lowest point. When I went back the next day. I remeasured the sag and it was at 17 ft. I had forgotten to compensate for the summer sun. The heat caused the wire to expand and created another 12 in. of sag. The lesson I learned that day is always to build above the minimum required height to maintain a safe clearance. For the spans I normally deal with, a good rule is to build at least 1 ft. above the minimum height.

The drip loop (installed by the utility) needs to be at least 10 ft. off the ground. To accommodate for the loop, the last clamp you install (always double-clamp the last attachment) needs to be around 13 ft. off the ground—assuming a level finished grade. The individual wires of the drip loop must be kept at least 3 ft. away from the bottom and sides of any operable window, stairs, balcony, deck, and so on. As long as the SE cable stays in its sheath, or the cable is in conduit, there is no minimum distance between it and any of the above items.

Connecting the meter base with the aerial utility

The four most common ways to connect the meter base to the aerial utility cable are as follows:

1. *Run style-U cable from the meter base and clamp it to the siding (no service head).* After installing the meter base, use a plumb line or level to draw a line straight up from the center of the meter base. The cable, as it leaves the meter base and clamps against the siding, must follow this line. You are allowed to arc the cable along the exterior wall or clamp it to the soffit and run it to another part of the wall. However, this looks ugly and should be avoided if at all possible.

To attach the cable, first screw the bottom half of the watertight seal into the meter base and slide the top half onto the SEC end. Next, slide in about 18 in. of cable—enough to connect to all three terminals inside the meter base. After inserting the cable, place the first clamp 12 in. above the meter base

Typical Gable-End Attachments

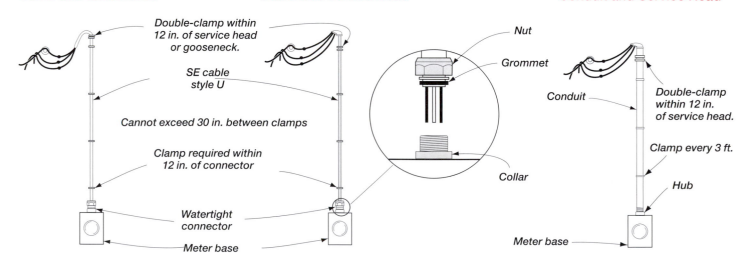

Cable and Gooseneck

Double-clamp within 12 in. of service head or gooseneck.

SE cable style U

Cannot exceed 30 in. between clamps

Clamp required within 12 in. of connector

Watertight connector

Meter base

Cable and Service Head

Nut

Grommet

Collar

Conduit and Service Head

Conduit

Double-clamp within 12 in. of service head.

Clamp every 3 ft.

Hub

Meter base

Minimum Clearances over Property

Service drop

10 ft. from bottom of drip loop (must increase to 12 ft. if cars have access)

12 ft.
Residential property and driveways

18 ft.
REX'S
Public streets

From Pole to House

NEC calls for minimum clearances on service drop. But ice and heat can cause aerial lines to sag below original construction height. To compensate for sag, construct above minimum requirements.

12-ft. minimum clearance

House

Lift pole

Barn

Adding a Lift Pole

When minimum clearances cannot be maintained across a span, you must add a lift pole.

Operable window

Violation of 3-ft. Rule

3 ft.

Individual strands within 3 ft. of operable window

XHHW

Around windows

If window is operable, you cannot install individual conductors (a cable such as SEU, which has a jacket, is exempt from this rule) within 3 ft. of the side and the bottom of the window, but overhead is okay. If the window is not operable, you can get as close as you want. The rule also applies to doors, porches, balconies, ladders, stairs, fire escapes, or similar locations.

and the others at 3-ft. intervals. (This is easier said than done: While working on a ladder, you must lift the heavy cable, place the clamp on the wall, and then screw it in place.) Double-clamp the cable where it bends down for the drip loop. (I use a cordless drill, a magnetic 5/16-in. hex driver, and 5/16-in. hex-head screws long enough to reach solid wood inside the wall. It helps if you have someone arc the cable for you as you attach the clamps.) Once the cable is up, tighten the cap on the watertight seal and add silicone where the cable enters the seal.

To splice the SE cable into the meter base, cut back the stranded neutral until it is the same length as all the strands and it fits under the neutral clamp. Cut the hot legs approximately 1 in. shorter than the neutral. Now all you have to do is bolt them down. Don't forget to smear on anticorrosion grease if you are using aluminum conductors. You do not have to ground the meter base. The neutral bus in the center of the meter base is already connected to the frame.

2. *Run style-U flat cable from the meter base and clamp it to the siding (with service head).* The installation method is basically the same, but in this case, you'll also need to deal with some type of service head at the top of the cable. Inside, the individual wires are stripped out of the cable and then point straight down to come out the bottom. They attach to nothing and the unit is not watertight. Their purpose is to provide an umbrella to cover the cable where the wires break out of the sheath; this keeps water from running down into the cable. If at all possible, I avoid using such devices on style-U SE cable because they are difficult to

The watertight seal needs some help to be watertight. Wrap Teflon on the threads, then slide on the grommet that fits into the collar and holds the cable. Finally, slide on the anti-grab metal collar and a screw cap.

Use silicone sealant where the cable enters the rubber seal. Without it, water will leak through the hub and into the meter base.

The most common meter base hookup, type-U aerial, should wind up looking like this.

This wired aerial service meter base uses individual cables.

install. However, if your local area requires it, you have no choice.

You'll need 3 ft. to 4 ft. of cable coming out of the service head for the utility to splice into and form a drip loop. Make sure you keep the sheath of the style U intact all the way into the service head; otherwise, the service head won't be able to do its job.

3. *Run steel conduit from the meter base, through the soffit, and up through the roof (service head mounted on top of a mast).* This method is used mostly on single-story houses where extra height is need for the utility clearance.

ABOVE CODE   For masts above the roof, I use only 2-in.-diameter or larger thick galvanized steel. Smaller pipe can't take the stress

of the utility cable pulling against it and may bend. Never use PVC conduit above the roof; it can bend and snap. Always double-clamp within 12 in. of the goose-neck or service head.

When calculating the length of steel pipe needed, allow for 3 ft. to 4 ft. above the roof. Above that, you may need to tie the upper end of the mast to keep it from being pulled over by the weight of the utility cable. The utility connects to the mast about 12 in. under the service head and places its attachment and drip loop around 2 ft. to 3 ft. above the roof (the minimum distance is 18 in.). Never have a splice (coupling) in the mast section of conduit above the soffit and roof. The threaded section of the metal is too

Mast head

Pulling pressure ⟵

Mast

Flange glued and screwed

2 ft. to 3 ft.

If mast height exceeds 4 ft., bracing should be considered. Also, no coupling allowed above roof.

18 in. minimum to point of attachment

Cut hole through soffit and roof with minimum gap around pipe.

Soffit

4 ft. maximum

Conduit clamps every 3 ft.

2-in. galvanized steel pipe threaded into meter base

Silicone at thread/hub interface

Meter base

weak to hold the stresses of a heavy utility cable.

The utility cable must not swing over any part of the roof other than the overhang. If it does, it places those who work on the roof, such as anyone cleaning gutters or repairing shingles, in serious danger. (The utility doesn't like running cables over the roof either, but it is worried about the cable, which would be ruined in case of a fire.) If the service entrance looks as if the cable will cross the roof, find a different route or move the location of the meter base. In some cases, the utility can place a drop pole to change the approach angle of the cable to the house.

Installing a service entrance through the soffit is not easy. To do this, drill a hole in the roof and soffit for the conduit to reach above the roof. Then carry a heavy 2-in.-diameter pipe up to the roof and slip it down through both holes. The threaded end screws into the hub of the meter base. Put some

400-Amp Service Entrance

When running cables larger than 4/0 aluminum, the installation gets tricky because of their thickness. Most of the time, it's easier to run two cables of the same diameter, length, and material in the conduit in parallel rather than run one large cable. (Never use one large-diameter cable and one small-diameter cable—all current would flow through the larger cable because of its lower resistance.) For example, for a 400-amp service, an electrician may choose to run two 4/0 cables to each leg of the meter base. In that case, it is imperative to mark each phase wire with colored tape. For example, use blue tape to mark the two wires connecting to the meter base's left terminal (mark the cable at the meter base and at the service head). The wires connecting to the meter base's right terminal will then need a different color, such as yellow, to identify them. The neutral is always white. Never use green—it is for ground. This way, the utility crew will not get the wires mixed up. Another way to prevent any mixup is to use two conduits of the same material, one for each cable group, as opposed to putting all cables into one large conduit.

400-Amp Service with One Conduit

400-Amp Service with Two Conduits

■ **WORK SAFE**
■ **WORK SMART**
■ **THINKING AHEAD**

The secret to cutting a round hole on a sloped roof is to use a template. Using a scrap piece of conduit, cut one end to match the slope of the roof, so that the pipe stands vertically when positioned on the roof. Trace along the base of the pipe to establish a cut line. When you make the cut, lose the line. You'll need the extra wiggle room to slide in the conduit.

pipe dope on the threads before you tighten it with a pipe wrench. Install the first clamp within 12 in. of the base and the others at 3-ft. to 4-ft. intervals. You will also need to install a weather seal at the pipe/roof interface. The hole cut for the conduit coming through the roof should be as close as possible to a perfect fit around the conduit/mast. If there is an excessive gap between the pipe and the wood, the pipe may bend over as the weight of the utility cable pulls it. This will make the roof seal leak.

Once the conduit is installed, cut the SE cable to length and slip it into the conduit either from the roof or up from the meter base. Add an extra 5 ft.—4 ft. for the drip loop and 1 ft. for the meter base. Next, insert the cables through the service head on the top of the mast and mount the service head. Let the SE cable hang down from the service head. The utility will form the drip loop as they splice in their utility cable. Lastly, splice the cables into the meter base, making sure the neutral is noted with white tape at the drip loop and at the meter base.

4. *Run schedule 80 PVC conduit up the siding and terminate it in a service head on the side of the house.* Running PVC conduit up the siding is fine as long as it does not extend through the soffit and is not used as a mast. Its light weight makes it easy to work with (especially when you're working alone), but it does not have the strength to support the weight of the utility cables.

To install PVC conduit, follow the same steps as described for steel (see p. 65). The cable lengths of the drip loop and the splice in the meter base are also the same.

There is, perhaps, a fifth way of bringing an aerial service into a residence—one that happened to be the worst I'd ever seen. The owner had laid the style-U SE cable on the ground from the meter base pole (farms back in the old days had the meter on a pole in the center of the farmyard) all the way to the house. Dogs had been chewing on the cable but as yet hadn't bitten through to the hot conductor. The cable then went in through a window (that always stayed open), through the bedroom, and across to a panel on the other side of the room. Although this method may be a fast way to install a service entrance system, and it could double as a handy clothesline, it's not one that I recommend.

Connecting the Meter Base to the Panel

Once I was working on a house that was about 10 years old. The owner was complaining about dimming lights and everything acting "funny." First, I checked the voltages at the meter base that was mounted on the outside corner. All the voltages checked fine. When I checked the voltages at the service panel, which was located in the basement somewhere around the middle of the house, the voltages were funny: There was low voltage (80 volts) from one hot leg to neutral and high voltage (160 volts) on the opposite hot leg to neutral. And the voltages were fluctuating. Upon closer examination, I discovered that the neutral was being eaten through by water that had accumulated inside the sheath of the service cable connecting the panel to the base. This is quite a common occurrence, and it is one of the many reasons why I won't separate the panel and the meter base.

The Art of Stripping Large Cable

Holding the utility knife parallel to the ground, slice into the insulation until you feel the blade on top of the conductors. Tilt the blade so that it slides over the conductors and cut the insulation to the end.

Circle the cable with the utility knife cutting three-quarters of the way through the insulation.

Using electrician's pliers, peel the insulation off the cable.

Use 9⅛-in.-long conduit to connect a meter base to a service panel through an 8-in. block wall (with a 1½-in.-thick wood spacer behind the panel). Placing only the male adapters back-to-back shortens the conduit length to 2½ in. Cutting ½ in. off each adapter hub further shortens the distance to 1½ in. If the meter base and service panel are closer than 1½ in., use a metal all-thread nipple.

Locating the meter base back-to-back with the service panel, or as close as is practical, simplifies the installation, reduces installation cost, and as indicated by the example above, makes for a better job. Remember, water will follow the cable any time the SE cable leaves the bottom of the meter base to enter the house. Meter bases are not watertight—they can leak through the "watertight" hubs on top or around the glass.

You can connect the meter base and the service panel in several ways:

1. *Via flat, all-in-one style-U cable.* If you use style-U cable, conduit is not always required between the meter base and the service panel. (Be sure to check local regulations.) However, if you don't use conduit, there's a risk that someone may

Use a PVC saw for cutting plastic conduit. Its stiff, wide blade keeps the cut straight.

drive a nail or screw into the cable. To make the interface, all you need is an NM connector in each box with a bushing over the threads. Run the cable between the two boxes.

2. *Via cable inside a plastic or metal conduit.* If you use individual cables, such as THHN/THWN, you must use some type of conduit to enclose the cable. I prefer plastic schedule 80 conduit rather than metal because it's easier to cut, easier to install, and less expensive. I also prefer running individual conductors rather than all-in-one (style-U) cable, because individual conductors are easier to bend and maneuver in the conduit and panels.

The length of conduit nipple for a back-to-back run from the meter base to the service panel varies. For a block wall with a treated 2×6 spacer mounted inside the building, the conduit length needs to be around 9⅛ in. or just slightly less (slightly less and you'll never know; slightly more and it won't fit). Putting two hubs against each other yields 2½ in. to 2⅝ in. (cut the pipe that fits into the hubs 2⅜ in. long). For panels that are literally back-to-back, use a metal all-thread nipple and a grounding bushing.

3. *Via 2-in. flexible liquid-tight conduit.* If you've got a lot of turns to make and you're required to encase the cable in conduit, then liquid-tight conduit is the way to go. The flexible conduit and fittings cost more than PVC, but they're faster to install because there's no gluing or angled fittings to deal with. I use liquid-tight because the smooth, fittingless conduit makes it extremely easy for me to pull through individual cables.

A grounding bushing screws onto the metal conduit that exits the meter base and goes into the house (bottom left). Note the ground wire (#4 copper for 200-amp service) connecting the bushing to the ground lug of the meter base (bottom center).

Connecting to the main panel in a basement

I don't recommend it, but if the main panel must be located in the basement, there are three ways to get the SE cable from the meter base to the panel. The first option is to run the SE cable (make sure it's rated to be outside and in direct sunlight) without conduit right into the basement. However, most codes don't allow you to run cable through concrete without some type of protective sleeve. To get around this, you can run the cable through the siding and rim joist immediately above the concrete and then turn down to the panel (some areas will still require conduit).

The most common method of running cable into the basement is to run schedule 80 PVC conduit immediately out of the bottom of the meter base, or cutoff panel, just far enough to get into the basement. The PVC normally travels straight down, then through a long sweep or a 90-degree—immediate-turn—elbow

Three Ways to Bring Cable into the Basement

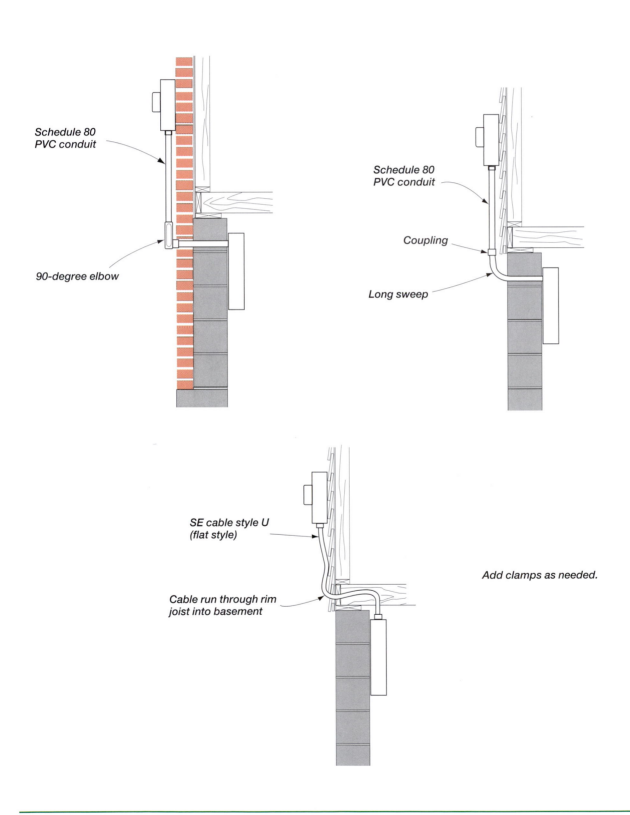

Schedule 80
PVC conduit

90-degree elbow

Schedule 80
PVC conduit

Coupling

Long sweep

SE cable style U
(flat style)

Cable run through rim
joist into basement

Add clamps as needed.

to go through the wall. (Remember to seal the hole around the conduit; otherwise, if the hole is below grade, the basement will flood with the first rainstorm.) Once through, you can either go directly into the main panel, or if you're still above, use an elbow to turn down.

Overall, installing the panel in the basement is not only expensive and time-consuming, it's also difficult to pull the individual conductors through the conduit, especially the elbows. In addition, water coming in around the meter itself (and sometimes through the alleged watertight connector on top of the meter base in aerial installations) will sometimes flow down the cables and into the main service panel, destroying the main breaker. I don't recommend putting the main panel in the basement, unless it's absolutely necessary. If you are going to do it anyway, run the cable or conduit into the bottom of the panel. Once in, bend the conductors up and route them along the panel gutter to the top of the panel to connect with the main breaker. This will, at least, keep water out of the main breaker.

Cutoff panels

Many locales require a cutoff panel immediately adjacent to the meter base if the two are separated by 5 ft. or more. (The distance depends on your local restrictions: Some locales allow 10 ft.; in my county, there is no set distance.) The purpose of the cutoff panel is to protect the cable between the meter base and the service panel. Without it, there would be no overcurrent protection for this cable as it travels through your house. A single nail or screw could cause a meltdown. By the time the fuse on the transformer kicks in, your house will be on fire. Expect to pay around $500 for this switch—another good reason to keep the meter base and panel back to back.

Liquid-tight conduit doesn't require additional fittings to change direction, and it has a slick interior surface.

To make tight bends in stiff cables, use the hole in the handle of a large adjustable wrench for extra leverage.

■ **WORK SAFE**
■ **WORK SMART**
■ **THINKING AHEAD**

When large-diameter SE cables bend up inside the panel, they can get in the way of smaller wires and even prevent you from screwing on the lid. To keep cables under control, pull them against the back of the panel with cable clamps and 5/16-in. hex-head screws.

CHAPTER 4

Panels and Subpanels

THE MAIN service panel and subpanel are the heart and soul of a home's electrical system. Through its outgoing wires flows the lifeblood of the electricity used throughout the residence. The main service panel has several important functions:

- It serves as a distribution center for all branch circuits.

- It houses all overcurrent devices that protect individual branch circuits.

- It houses a master switch that you can manually throw to cut off all outgoing electrical power from the service panel. (Some locales require the master cutoff switch to be adjacent to the meter base. If that is the case, there may or may not be an additional master switch in the panel itself.) The master switch also monitors the total amount of current the house uses and automatically cuts the power if it exceeds a specific amount.

Many installations fail, regardless of the class of the house, because some installers incorrectly assume that all main panels are designed the same way and therefore install the cheapest one they can find. The fact is, all panels are not created equal: There are good designs and there are bad designs. It's also important to realize that even the finest panel is no better than the person who installs it. A good panel gives you only the potential of a good installation—it doesn't guarantee it. The installer must know the basics of a good installation and what pitfalls to avoid, as well as have the willingness to do the job correctly. In this chapter, I will help you pick and install a high-quality panel.

Elements of the Main Panel

The main service panel (also called the box, panel board, or load center) has many sections, and each one performs a specific function. The three most important sections of the main panel are the main breaker, the hot bus, and the neutral/grounding bus. To

Main Service Panel

Knockouts for cable entrances; also located on top, bottom, and sides of panel.

Gutter post

Main breaker

Gutter

Hot bus, or leg

Keep all branch-circuit wiring outside of gutter posts.

Insulated hot wires from the meter base attach here.

Utility neutral/ground cable attaches here.

Hot bus, or leg

Gutter

Neutral/grounding bus

Bonding strap, or panel bond

Gutter post

understand how the panel works, you must first understand what each section does.

Main breaker

The main breaker, also called the main disconnect switch or just "the main," is normally located at the top of the main panel. All the power that comes into the house goes through this switch. All adults in the home should know what this switch does and where it's located. If there is ever an electrical emergency in the house, or perhaps a fire, throwing this one switch will allow you to cut off *all* power.

Although it can be used as a manual disconnect, the main breaker's primary purpose is overcurrent protection. It monitors the total amount of current coming into the house so that it cannot exceed the current-carrying capacities of the SE cable, meter base, and main panel. If this amount is exceeded, the main breaker disconnects the incoming power from the buses. This amount, in amps, is written on the breaker handle.

The two hex-screw terminals immediately above the main are for the two hot SE cables coming in from the meter base. The two plastic pieces immediately above the main and below the buses are called gutter posts. All branch-circuit wiring should remain outside the posts (wires cannot cross over the breakers from one side of the panel to the other).

Hot bus

The main breaker transfers power to the hot bus. The hot bus consists of two copper or aluminum strips, which are sometimes called legs; they are located immediately below the main and run down the center of the panel. Each hot leg has a row of tabs that allow for the insertion of circuit breakers. The bus takes the hot power and distributes it to the circuit breakers.

Each leg is the same voltage, in reference to neutral, but acts as an independent power source. Simply put, the current in the hot bus comes to each leg from alternate sides of the utility transformer. When measured from leg to leg, the voltages add up to 240 volts; the voltage for each leg, to ground or neutral, is 120 volts.

Neutral/grounding bus

The neutral/grounding bus provides a common return point for the current after the load has used it. It also provides a ground reference point. The neutral bus consists of two long aluminum strips with many screws; they are usually located on the outer edges of the hot buses. In a main service panel, the screws serve as attachment points for the neutral and ground wires (there will be a lot of them). In a sub-panel, the neutrals and grounds are connected to independent buses—one grounding bus and one neutral bus. The large hex screw at the top, called the main neutral lug, connects to the neutral coming in from the utility meter base.

Even though the neutral is grounded at the utility pole, an additional ground is provided by the installer at the main panel. This is the ground-rod connection and it is normally a large-diameter, solid-copper wire (4 or 6 gauge). For additional protection, a bonding screw is installed, connecting the neutral/grounding bus to the metal frame of the main panel. The bonding screw, called the panel bond, places the metal of the panel at ground potential so that it can never become a conductor if a hot wire touches it. In addition, all equipment

Main Breaker Load Center

200A

40 - 1" Spaces / 40 Circuits Max.

Series F
Quantity: 1
4104084 Rev. A

Siemens Energy & Automation, Inc. / Alpharetta, Georgia U.S.A.

7 83643 16501 4

UL

Actual Product Shown

INSTA WIRE™ **Speeds Your Install With Every Turn!**

- Indoor Enclosure (NEMA Type 1)
- 1 Phase, 3 Wire, 120/240V AC
- Combo Trim
- Copper Bus Bars

NOV 2 7

Catalog Number

G4040MB1200CU

grounding wires from every receptacle and every appliance, as well as all neutral wires (the white ones), connect to the neutral/grounding bus.

A well-designed panel has a bar that connects the two buses as one electrical unit. If the panel with this setup needs to be converted to a subpanel, which has its neutral isolated from the ground, the two can be separated easily by removing the bar. Not every panel has this handy option.

Picking a Panel

Before purchasing a main panel, you will need to know—or at least have a rough idea of—the number of circuits required for the house. Almost all houses can get by with 40 circuits for a 200-amp system, 60 for a 300-amp system, and 80 for a 400-amp system. If you need more than that per system, consider installing a subpanel.

The way to tell panels apart when buying them (other than by a part number or amperage) is by their circuit-quantity designation. That designation appears on the packaging as two numbers, such as 20/30 or 40/40. The first number indicates the total number of full-size breakers that the panel can hold if only full-size breakers are installed. The second number indicates the total number of circuits allowed in the panel. For example,

a 40/40 panel can hold 40 circuits and all 40 are full-sized—no dual breakers.

If the second number is larger than the first, then dual breakers are allowed. For example, a 20/30 panel can hold 20 full-size breakers for 20 circuits. Utilizing the same 20 tabs, 30 circuits can be obtained if the dual breakers are used at specific places within the panel. These types of panels normally have special tabs for the dual breakers located on the bottom of the buses.

It is possible to cheat and overfill the panel by installing dual breakers where they don't belong, which you may be tempted to do if you bought the smaller panel to save some money. Experienced electricians have learned how to break the metal "stop" out of the back of the dual breaker to allow it to fit on a full tab. This makes it physically possible to put 40 breakers in a 20-circuit panel or even 80 breakers in a 40-circuit panel. Don't do this. You not only void the warranty of the breaker and panel but you exceed the maximum number of breakers for which the panel is designed, creating a potential hazard.

Bus design

When you go shopping for a service panel, you will have a choice of three bus designs:

- Panels with all solid bus tabs, called full-size panels.

Check the specs on the box. The left side gives you its maximum rating (200 amps) and indicates that it can hold 40 full-size 1-in. breakers. The right side lists the part number and indicates that this panel is for interior use, is designed for 120/240 single phase, and has a copper bus.

Panels differ in their hot-bus tabs. The solid-tab panel (above) accepts only full-size breakers, while the slotted-tab panel (below) takes both full-size breakers and two-in-one breakers. Contrary to what you might think, the solid bus tab is the one you normally want.

■ Panels with all slotted (or split) tabs.

■ Panels with both solid bus and slotted tabs.

The advantage of full-size panels over split-tab panels is their size. A full-size panel has plenty of room for wiring and can only house full-size breakers, which I prefer. Split-tab panels can fit both full-size and two-in-one breakers. This may seem good at first, but having more breakers with less room can turn the panel into an overcrowded mess that is hard to work in.

In most cases, contractors use these smaller panels to save a few bucks. Ironically, I've found these panels even in million-dollar homes. There are times where a few extra dollars becomes money well spent, and buying full-size panels is one of those times. The one time when I may not use a full-size 40/40 panel is when I am installing a 300- or 400-amp system. The reason is that if I plan to put several high-amperage circuits in the second panel—for example, two heat pumps—I may need only eight slots to max out the panel's amperage.

ABOVE CODE In most cases, I use full-size panels. A larger panel is easier to wire and looks neater when you're finished. And in electrical work, neatness counts. When you finish a wiring job, no matter how well you did the design or laid out the wires inside the walls, the only part that shows is the panel. When an inspector sees a messy panel, a red flag goes up in his or her mind. Based on the appearance of the box, the inspector may start wondering about the overall quality of your work. Odds are good that he or she will spend much more time inspecting the rest of the house.

Panel size comparison. Both panels hold the same number of breakers, but the full-size panel on the left gives you much more wiring room than the split-tab panel on the right.

This undersized panel was filled the day it was installed. (Those two blank spaces over to the right are not empty slots—they're fakes.) The owner now has to install a subpanel or upgrade his service to wire in his new spa.

Mounting a Panel

Where you put the panel makes a big difference in its life expectancy. If it's at all possible, try to place the panel back-to-back with the meter base. A garage, for example, would be a perfect location: meter base outside, panel inside immediately behind it. Never install a panel in a bathroom or wet basement, because the moisture destroys it. To prevent other moisture-related damage, avoid placing water and drain lines directly above the panel and make sure that clothes dryers are properly vented outside.

Panels are maintainable items. For that reason, you must make sure that you are able to comfortably stand in front of the panel in order to work. Thus, there is a required minimum distance around the

The end terminals of this breaker had corroded so badly due to moisture damage that I had to cut the cable when the breaker died.

Panel Shopping List

All panels are not created equal. Some designs are great; others I would never use. What you want is an intelligently designed panel—one that is safe and wires fast, easily, and logically. Everything about the Siemens panel, shown here, suggests that an electrician designed it. Here is a shopping list of features that you'll want your panel to have:

■ A fall-safe lid. With some panel lids, an electrician needs three hands—one to hold the driver and two to keep the lid from falling and smashing his toes. With a Siemens panel, it's okay to have only two hands. The lid has two metal tabs that ride on top of the box frame.

■ A comfortable cutoff. I have made service calls where the only problem was that the homeowner couldn't get the breaker back on or off. This panel breaker has a wide handle that turns on and off vertically (side to side is harder to operate). You can use four fingers to pull it down and the palm of your hand to push it back up.

■ A movable neutral lug. Being able to move the lug to where the cable enters the panel speeds installation, leaves much more room in the box for the branch-circuit wires, and eliminates having to run the large utility neutral all around the inside of the box. To move it, first remove the setscrew from the lug. Look inside and you will see a Torx-head screw holding the lug body onto the bus.

■ Copper hot buses. Copper buses beat aluminum buses every time. Copper conducts electricity better than aluminum and has fewer corrosion problems.

■ Lots of neutral screws in the right location—immediately next to each row of breakers. Note that they are neither against the panel side where they'd block the knockouts, nor at the panel bottom or top, but right where they belong— next to the breakers. Cutting all the wires to the same length saves time and makes the panel look neater.

■ A system that allows several ground wires under one bus screw. Siemens buses are UL approved for up to three 12- or 14-gauge ground wires under one screw. Many manufacturers make you attach more than one wire per bus screw whether the panel is approved for it or not. You cannot, however, put a neutral and ground under one screw or two neutrals under one screw.

■ Special neutral bus attachment system. These patented square/standard drive screw heads speed up the wire attachment time dramatically because you can tighten the screws with a common 5/16-in. flat-blade screwdriver or a power driver and square drive bit.

■ Instant conversion from the main service panel to the subpanel. A Siemens main service panel has a tie bar at the panel bottom that connects the two neutral/grounding buses. Remove this tie bar, and the panel is instantly converted to a subpanel. Use one bus for the grounds and the other for the neutrals.

This bus screw is designed to secure up to three ground wires.

The back view of a panel lid. These metal tabs grab onto the panel frame to keep the lid from falling when the screws are withdrawn from the lid.

A Torx screw is located under the utility neutral lug. Remove it to move the lug body.

Clearances at the Panel

78 in.

30 in.

36 in.

Opening the Plates

Drill two ⅞-in. holes 6 in. apart, then open the slot between with a reciprocating saw on each side.

Run cable lubricant around the cut.

Double top plate

Main panel

Bottom plate

Run cable lubricant around cut.

Detail of Slot

6 in.

6 in.

2 in.

panel that must be kept at all times, and the door must be able to open at least 90 degrees. Because you may need to get to the panel fast in an emergency, you don't want to be climbing over boxes or tables to kill the power. Most common problems are due to renovations made after installation, such as new overhead plumbing, counters mounted in front of the panel, shelving mounted across the front of the lid, and appliances (especially the clothes washer) set in front of the panel. One home I worked on had a spa mounted in front of the panel—you had to stand in the water to work in the panel. Another had the panel mounted over a stove, and you had to crawl over the burners to get to the breakers.

Mounting options

Service panels are designed to be mounted several ways. If there is a solid backing, use the five small, prepunched holes in the back of the panel. First, hang the panel like a picture frame using the elongated hole in the top center of the panel. Next, level the panel, then install the four corner screws. If the solid backing is masonry, be sure there is at least a ¼-in. gap between the metal of the panel back and the masonry itself. The wood spacer must be pressure-treated. I normally use a treated 2×6 (be sure to get one that isn't twisted). If you need to protect the wires, build a vertical box, or plenum, to house the cables once they are installed.

If mounted inside the wall, the panel must be attached by at least two of its four sides. If the panel is mounted between two 16-in.-on-center studs, you'll lose the ability to draw cable through the sides of the panel unless you cut the wood. When mounting a panel between 24-in. on-center studs, attach one side of the panel to the stud and run a crosspiece above or below the box for the second side. Remember to leave enough of the

Removing Panel Knockouts

NM connectors are used whenever you're installing NM cable to a panel, a subpanel, or metal receptacles. Without the protection of the connector, the cable can be cut on a sharp metal edge or accidentally pulled out. Anyone can pop out the easy knockouts with a screwdriver, but what do you do about the stubborn ones?

Make sure you're opening the knockout in the direction it wants to open. Use a sharpened screwdriver to cut *sideways* across the attachment point.

Use a drill and a metal cutting bit to cut through the attachment point.

Use a fine-toothed sawblade in a jigsaw or a mini reciprocating saw (as I have here) to cut the attachment point.

panel sticking out from the studs so it will be flush with the finished wall.

> **ABOVE CODE** Use screws, not nails, to mount the panel. Nails work loose over time and do not allow for minor adjustments. I always use 5/16-in. hex-head screws.

If the box is mounted inside a stud wall, you need a way to get all the cables through the top and/or bottom plate within the wall cavity (for cables going into the attic or basement). You can drill holes for each cable or group of cables, but that takes time and is both ugly and messy. Instead, create two large oval slots next to each other, as shown in the bottom drawing on p. 82. Use the left slot for the cables entering the left side of the panel and the right slot for the cables on the right. Drill two 1-in. holes about 6 in. apart, then use a reciprocating saw to make a large oval cutout between the two. Coating each opening with some cable lube or liquid soap will make it easier to pull cable.

Roughing in wiring to the panel

After mounting the panel, rough in the cables to the panel location. If you are bringing the cables into a wall cavity with the two slots cut into the plate, bring the 120-volt circuits into the left slot and the 240-volt circuits into the right slot.

Do not bring the cables inside the panel at this time. To save time, most electricians run all the cables through the house and leave the cut ends dangling around the panel with the cable's destination written on its sheath. Once all the cables are run, the electrician then ties them into the panel. Working this way is faster than wiring the cables one by one, but it forces you to sort through 30 to 40 cables. You can try searching for cables one by one or pick the first cable that's

handy and bring it in with no organization at all. However, there is a better way.

> **ABOVE CODE** During the rough-in stage, write where the circuit goes on the sheath and wrap the end of the cable with colored tape: for example, yellow for the kitchen circuits, red for the lighting, blue for the bath, and green for the receptacles. Leave the 240-volt circuits as is. Color-coordinating the circuits makes it easy to identify them by group and bring them into the panel in an organized fashion.

Cable Management

The rule for cable managment is to be consistent and to have a system. You won't wire every panel exactly the same, but you can wire every panel in the same basic way. To keep things organized, I bring all the 120-volt circuits into the left side of the panel and the 240-volt circuits into the right. I further break down the left side into groups. I first wire the kitchen circuits into the upper left of the panel. Then, working downward, I wire the lighting circuits, then the bath, then the receptacle circuits. I like to wire the receptacle circuits so that the washer/ utility room receptacle is the last breaker on the left side. Then I put the electric dryer receptacle opposite it, on the 240-volt side. Of course, you don't have to use my system, but make sure you have one.

Be just as consistent when bringing cables into the panel. That is, bring the cables that go to the left-hand circuits into the left side of panel and the ones that go to the right-hand circuits into the right side of the panel. Do not crisscross the cables once they are in the panel— it looks messy and wastes valuable space.

Filling a Panel

Shown here is a typical panel fill for an all-electric house. I have all 120-volt circuits on the left side of the service panel and have broken them down into four sections: kitchen, lights, bath, and receptacles. All 240-volt circuits are on the right. (The numbers listed below do not coincide with the numbers on the panel—they just show you how many breakers it takes.) As you can see in the list, even a large 40-circuit panel can fill up fast. Each house, of course, is different. There are also some appliances that I haven't listed, such as a freezer, sewage pump, sewage pump alarm, whirlpool, and central vacuum, each of which takes its own dedicated line. When your box starts filling up, as this one has, consider installing a 300- or 400-amp system to have more breakers.

Service panel

Kitchen

Lights

Bath

Receptacles and AFCIs

120-volt single-pole circuits

240-volt double-pole circuits

Left Side of Panel

(numbers do not correspond with panel numbering)

1. (20 amp) Dishwasher
2. (20 amp) Garbage disposal and instant-heat water dispenser
3. (20 amp) GFCI kitchen counter small appliance #1
4. (20 amp) GFCI kitchen counter small appliance #2
5. (20 amp) Refrigerator (optional)*
6. (20 amp) Kitchen, dining room, and pantry wall receptacles**

—

1. (15 amp) Smoke alarms, all 3-way switched lights***
2. (20 amp) Lights
3. (20 amp) Lights
4. (20 amp) Lights
5. Spare
6. Spare

—

1. Bath—GFCI
2. Bath #2 GFCI or second receptacle in first bath

—

1. Garage receptacles—GFCI
2. Unfinished basement receptacles—GFCI
3. Outside receptacles—GFCI
4. Bedroom #1 and #2 branch circuit—AFCI
5. Bedroom #3 and #4 branch circuit—AFCI
6. Utility room (clothes washer receptacle)

Right Side of Panel

(numbers do not correspond with panel numbering)

1 & 2. Electric stove
3 & 4. Electric water heater
5 & 6. Heat pump outside unit
7 & 8. Heat pump inside unit
9 & 10. Bath #1 heater
11 & 12. Bath #2 heater
13 & 14. Water pump (for us country folk)
15 & 16. Spa GFCI (we country folk have these, too.)
17 & 18. Welder, table saw, or kiln (for do-it-yourself folk)
19 & 20. Electric dryer

* Code allows this to be a 15-amp circuit. I prefer to keep it a 20-amp.
** The NEC allows these receptacles to be on the kitchen counter circuits. I do not. My Above Code system requires that their loads be taken off the counter circuits to allow more available load for the counter circuits.
*** Here we are putting together all the 14-gauge loads that use 3-conductor cable. The 14 gauge can be used here because there is either no current on the load (smoke alarms) or minimal current (3-way switch loads). All outlets that enter the bedroom (lights, receptacles, smoke alarms, etc.) require AFCI protection.

Bringing Cables into the Panel

Once all the cables are run, it's time to bring them into the panel. Regardless of the panel type, all cables must enter the box via circular knockouts that use NM connectors or conduit to keep the wires and cables from getting cut on the knockout's sharp edges. Any unused holes that got knocked out by accident must be plugged.

In case you're wondering, you are allowed to put more than one cable through an NM connector. Typically, the common ½-in. connector can take two 12- or 14-gauge NM cables. If the cable is #4 gauge or higher, the lip of the connector or conduit needs a bushing to protect the cable's insulation from abrasion. Never bend any wire or cable into a sharp 90-degree angle once it gets through the connector or you'll damage it.

The next problem is identifying circuits once you strip off the sheath. If you want, you can leave about 1 in. of sheath intact on the cable after it enters the panel so that you can write on it where it goes. There are fancy numbered sticker systems, but they don't work any better than freezer tape. Of course, you don't have to identify the circuit if you can remember where it went once you strip off the sheath. But if you have to leave in the middle of the job, you may forget.

To route the wires to the breakers, run them along the outside edge of the panel frame until the wire gets close to its breaker, then make a gradual bend over to the breaker. Do not make sharp 90-degree bends—this can damage the wire by creating a stretch area on the outside of the bend where the metal pulls apart, as well as a hot spot on the inside of the bend where the metal compresses and overheats. Install the neutral wire in the neutral bus screw closest to the

breaker at which its accompanying hot wire terminates. Do not combine two neutrals under one screw and do not put a neutral with a ground wire.

In the old days, the white wires of a 120-volt circuit always went to the neutral bus, and its accompanying hot wire (normally black) always went to the breaker or fuse. Not anymore. With the advent of GFCIs and AFCIs, some of the neutrals go to the breakers as well. It is imperative that you read the side of these breakers to determine which wire goes where before the breaker is installed in the panel; otherwise, you may reverse the neutral and load connections on the breaker. Once the breaker is installed, you can't read the printing on its side that tells you which wire goes where.

On 240-volt circuits, both the hot and the white wire go to the breaker. In this case, you need to put a piece of black tape on the white wire to show that the wire is not a neutral but a hot conductor of a 240-volt circuit.

Panel bonding

Bonding refers to connecting two conductors to form a complete current path—as in bonding the grounding bus to the metal frame of the panel. A bonding jumper or screw normally makes the connection. Once the bond is in place, if a hot wire touches the panel frame, the fault current flows through the bond jumper to the bus, allowing the breaker to kick. If the panel bond is not present, a complete circuit for current flow does not exist and the breaker will not kick. In that case, the frame stays hot and can electrocute someone.

In addition to the panel bond, any metal conduit extending from the service panel into the meter base must be bonded to the panel (or meter base) in case a hot wire faults against the conduit. This is normally done via a grounding bushing that screws over the sharp threaded end of the conduit (see the photo on p. 89). The bushing grounds the conduit and protects the cable against the sharp edge of the conduit's threads. (I'll talk more about earth ground connection, water line ground, and gas line bonding in chapter 5.)

Subpanels

A subpanel's most common use is to provide a secondary panel some distance from the main service panel. For example, if you have a long ranch house with the kitchen, baths, and utility rooms on the opposite side of the house from the service panel, you will need to run a couple dozen or more cables from one side of the house to the other. A better option is to run one large cable and terminate it in a subpanel

The bonding screw grounds the panel so that it can never become a conductor if a hot wire touches it.

Neutral Current Can Be Lethal

Even though neutrals are not considered a "hot" conductor, the same amount of current that flows through the hot wire flows through the neutral wire. If the 120-volt load draws 10 amps of current, you have 10 amps leaving the breaker on the black hot wire going to the load. After it goes through the load, there will be 10 amps returning to the panel on the white neutral. Either wire can kill you if you put yourself in series with the circuit.

Picking and Installing NM Connectors

NM connectors such as these are inserted into the knockout holes of the panels.

To install a connector, use a screwdriver and knock the center out of the prepunched knockout. Some knock-outs can be bent in only one direction.

Insert the NM connector through the panel, then tap-tighten it with a screwdriver and hammer. Orient the screw heads of the connector so they are facing forward.

Metal or plastic bushings are required on any sharp-edged thread extending into the panel and on any conduit carrying #4 or larger cable.

adjacent to those rooms. In this case, the subpanel saves labor and material costs.

Occasionally, subpanels are wired immediately next to a main service panel to solve circuit-shortage problems. This happens when the original panel is too small (probably a low-bid item) and the house needs more circuits (not necessarily more amperage) than the panel can supply. Some areas require a cutoff panel adjacent to the meter base on the exterior wall. If that is done, then the main service panel becomes a subpanel because it is downstream from the cutoff panel.

ABOVE CODE You are not required to put a main breaker within a subpanel, because the feeder cable from the main service panel or cutoff switch is always protected by the breaker at the main panel. However, I recommend installing a main breaker in a subpanel. Without a one-switch-controls-all breaker within the subpanel, you will have to go to the main panel to kill the power. If the panel that houses the cutoff breaker is on a different floor, at a distant location, or perhaps outside in the rain, you might get lazy and not turn off the power to the subpanel, thereby taking a few unnecessary risks. In the event of an emergency, cutting off the power at a nearby subpanel may save a life.

Wiring a subpanel

Subpanels require a four-conductor cable: two insulated hot conductors, one insulated neutral, and one ground. The two insulated hot wires provide the 120/240-volt power, and the insulated neutral return provides the return path for all the 120-volt circuits. You don't need a neutral return for the 240-volt

A grounding bushing grounds the steel conduit that enters the panel. This is required on any metal conduit between the meter base and the service panel.

circuits. Since 240-volt circuits already have two wires, that's all that's needed for current to flow. Current flows from one wire, goes through the load, and comes back via the other.

In any subpanel, grounds and neutrals must be kept separate from each other. The ground wires must go to a bus that is bonded to the panel's metal frame. The neutrals must be on a bus that is isolated from the frame and all ground wires. If the grounds and neutrals are intermixed, you will have neutral current flowing on all the ground wires and everything that is grounded, such as water lines and ductwork.

People sometimes ask why it is required to separate grounds and neutrals at the subpanel and not at the main panel. After all, if current can flow on the grounding system at the subpanel, then why can't it flow at the main panel, too? The answer is that current is always trying to get back to where it came

WORK SAFE
WORK SMART
THINKING AHEAD

Long sheet-metal screws are frequently used to replace lost blunt-end panel screws. Don't do it. If the sharp-tipped screw punctures the SEC cable, you may not live to regret it.

from—the utility meter. In the main service panel the current returning from the neutral conductor wants to go to one place—the transformer. The only way it can do that is via the SE utility neutral cable. It can't go back there via a house ground wire; thus, no current flows on the grounding circuit.

The current at the subpanel also wants to get back to the utility transformer. But in this case, the subpanel is a distance away from the utility transformer. If we were to tie the grounds onto the neutral bus at the subpanel, the neutral current could also flow on the ground wires to get back to the main panel (as well as the utility neutral), because they also terminate at the same point in the main service panel.

If the subpanel is in an exterior building, such as a shop, add a couple ground rods to the system as well. That is, ground the subpanel just like you would a main service panel—just make sure you tie it into the ground bus, not the neutral bus. This will bleed off any stray electromagnetic (EMF) voltage that may build up on the metal frames of the shop equipment (especially during thunderstorms). It's true that the stray voltage would bleed off via the ground wire going back to the main system ground at the service panel, but that route takes you through a lot of splices that have resistance. This is especially important if you are using any surge arrestors. It's better to keep surges out of the house, rather than send them to it.

Wiring a Subpanel

In a subpanel, the ground and neutral buses are separated to keep the neutral current off the grounding system. A main breaker is not required.

Main breaker

Neutral/grounding bus

SER cable

Ground wires to ground bus

Double-pole breaker

Neutral wire to neutral bus

Hot

10-3 w/g cable

Both the neutral and ground wires of the outgoing cable can connect to the same bus.

Panel bond

Neutral bus

Grounding bus

To ground rod

Tie bar connects both buses.

Main Panel

Subpanel

Converting a Main Panel to Subpanel

Main Panel

Two insulated hot wires

SEU (three wires required)

Main breaker required

Bare neutral

Panel bond on
either bus

Neutral and
grounding buses
are connected
with tie bar.

> ■ **WORK SAFE**
> ■ **WORK SMART**
> ■ **THINKING AHEAD**
>
> If you are using a screw as a panel bond and it is among a number of other screws and hard to see, paint the head green so the inspector will have no problem finding it.

Main Panel Wired as Subpanel

Two insulated hot wires

SER (four wires required)

Ground wire (bare or insulated)

One insulated
neutral

Main breaker optional

Grounding bus

Neutral bus

Panel bond on
grounding bus only

To make a subpanel from a main panel, isolate the grounding bus from the neutral bus. Some panels allow you to do this by removing the tie bar connecting buses.

Neutral Current Flow

When neutral and ground wires connect on a subpanel, subpanel neutral current flows along all grounded conductors and any metal objects that touch them, and then back to the service panel. This route is parallel with the neutral current that flows back via the neutral conductor connecting the two panels.

To the utility transformer

Neutral that connects to the service panel

Neutral current running back to the utility transformer via SEC

Neutral current running along the water heater ground wire to the metal of the water heater jacket

Neutral that connects to the subpanel

Neutral

Main panel

Neutral current running the metal water heater jacket to the metal water line

Water heater ground wire attached to the neutral bus

Neutral return current splitting because the current sees two paths back to main panel

Water heater

Neutral current running the water line ground wire

Neutral current splitting off because it sees another route back to the main service panel through the grounding system

Subpanel

Neutral return current from any 120-volt load

Neutral current running the metal water line

Water line tap

Metal water lines

Water line tap

Ductwork

Metal water line touching metal ductwork and neutral current transferring to water line tap

Metal water line touching metal ductwork and neutral current transferring to ductwork

Sizing the subpanel feeder cable

If the subpanel is in the service-entrance system (a meter base, a cutoff panel next to the meter base, and a breaker panel, which is now a subpanel), the cables feeding both panels are sized as SE cables. If there is a cutoff panel downstream from the main service panel, size the cable that feeds it the same as you would for a utility service cable for amperages of 100 and above.

If the subpanel feeds off the main service, as in the long ranch house example above, still use the SE cable sizing chart for currents 100 amps and above. For amperages less than that, follow Table 310-16 of the NEC. For example, if the subpanel is fed via a 30-amp breaker from the main panel, use 10-3 with ground NMB. For a 100-amp feed, you can use #4 copper cable. Always remember, the feeder cable that feeds a subpanel must be a four-conductor cable (hot, hot, insulated neutral, and ground).

ABOVE CODE If possible, always use a copper conductor to feed the subpanel. Copper is a better conductor, which means that you can run a smaller cable, and that makes for an easier pull. Also, copper doesn't corrode as readily as aluminum.

Balancing the Load

In theory, and as much as is practical, the load needs to be balanced as it leaves the panel and enters the service-entrance wires. A balanced load is wired in the panel so that the current on one 240-volt leg is equal to the current on the other; this way, the current cancels out in the neutral. Balancing requires you to analyze the circuits to determine the load they will serve. After that, you must employ some common sense.

Balancing lowers the heat buildup in the wire terminals and allows you to put more load on the panel. Let's assume you have a 200-amp panel and main breaker. If you wired the load so that 300 amps of current is pulled through only one leg, the main would trip. However, if you split, or balanced, the load, the current would be reduced to 150 amps on each leg—well within the 200-amp rating of the main.

In most main panels, the tabs on each leg of the hot bus are arranged in opposite phases, so balancing occurs automatically. As single-pole breakers (those that use only one hot leg of the panel) are installed down one side of the panel, they automatically get put on alternate phases because of the tab/bus design. If a double-pole breaker (one that uses both hot legs of the panel) is installed, the current will automatically balance because the breaker, by design, fits on opposite phases.

Inside the house, if two loads of any kind will be on at the same time, put them on opposite phases. For example, if two circuits—one with a freezer pulling 5 amps and one with a refrigerator pulling 5 amps—are put on the same phase, their current on both the neutral and the service-entrance hot leg will add up to 10 amps. If the same circuits are put on opposite phases, 5 amps will flow through each hot leg and will cancel out to 0 amps in the neutral. This is why the neutral is allowed to be smaller than the two hot wires. Even if the neutral load doesn't cancel, it will be reduced by the amount of the smallest load. That is, if one load pulls 10 amps and the other 6, the resultant current flow through the service-entrance neutral will be 4 amps if the loads are on opposite phases, 16 if they're not.

■ WORK SAFE
■ WORK SMART
■ **THINKING** AHEAD

Never paint or put wallpaper over the cover of the panel. The panel should be conspicuous in case someone needs to turn off the power fast.

Balancing the Load

A balanced load is wired so that current on one leg is equal (or close) to current on the other, so current cancels out in neutral. If two circuits—one with a freezer pulling 5 amps and one with a refrigerator pulling 5 amps—are put on the same phase, or leg, current in both the neutral and the service-entrance hot leg will add up to 10 amps. If the same circuits are put on opposite phases, they will still add up to 10 amps in the hot leg but will cancel out to 0 amps in neutral.

Unbalanced Load

0 amps

10 amps

All loads on the same bus are added in the neutral wire and the leg, so neutral will be carrying 10 amps.

Neutral
Hot
To freezer: 5-amp load

Hot
Neutral
To refrigerator: 5-amp load

10 amps will flow through this bus— 5 amps for a freezer and 5 amps for a refrigerator.

Balanced Load

5 amps

5 amps

Loads installed on opposite buses split amperage and cancel in neutral, so 0 amps flows through.

Neutral
Hot
To freezer: 5-amp load

Hot
Neutral
To refrigerator: 5-amp load

5 amps will flow through this bus for the refrigerator.

5 amps will flow through this bus for the freezer.

■ **WORK SAFE**
■ **WORK SMART**
■ **THINKING AHEAD**

When removing an old paint-encrusted service panel, run a utility knife around the edge of the lid to break the paint-to-wall seal.

If you're building a shop with a 240-volt feed and two 120-volt machines will be running at once, put them on opposite phases so that the current will cancel in the service-entrance neutral cable and in the neutral wire leading to the shop. If the shop only has a 120-volt feed, there is nothing you can do. It takes two opposing currents to cancel. With only one current, nothing will cancel.

CHAPTER 5

The Art of Grounding

ROUNDING IS the foundation of the electrical system. An improperly grounded electrical system is a hazard to both life and property; it can damage or destroy appliances and even kill those who use or come in contact with them. One of my customers is alive today only because of a grounded flue pipe. At my insistence, he grounded it upon installation. During an extremely violent thunderstorm, a ball of lightning came out of an outlet and flew across the room as if to hit him straight between the eyes. At the last minute it changed course and diverted into the grounded flue pipe. It is imperative that residences, especially those with computers, answering machines, and other high-tech sensitive equipment, have an extremely good grounding system. Otherwise, these expensive machines will wind up in the high-tech junk pile.

In this chapter, I explain how to install a proper grounding system—not just a code grounding system, but also my Above Code grounding system. The fact is, the NEC doesn't care how high the ground resistance is as long as at least two rods are driven. If there's a high ground resistance, a surge won't be diverted into ground but, instead, into that delicate electronic gear mentioned above. In comparison, my system gives you the lowest possible ground resistance, keeping both life and property as safe as possible.

Be selective when purchasing grounding material and take your time installing it—one wrong piece of equipment or one installation mistake can cause the entire system to fail. A grounding system is like a string of old-style Christmas tree lights. All the bulbs, but in this case connections, are in series. If any one is loose, burnt out, corroded, or installed incorrectly, the system won't work.

Definition of a Grounding System

Grounding is a process designed to keep all noncurrent-carrying conductors at a 0-volt potential. (A noncurrent-carrying conductor is a material that *can* conduct electricity but normally does not, such as the metal frame of a washer or dryer or the metal housing of an electric drill.)

Typical Ground-Rod System

Ground rods connect the electrical grounding system to the earth. A minimum of two rods may be required; however, more is better, especially in lightning-prone areas.

The location of the ground rods can lower ground resistance. At a gutter downspout, rainwater will saturate the ground around a rod, lowering ground resistance.

Gutter

Main panel inside residence

Downspout

Grounding wire

Ground rods

Moist soil

Minimum 6 ft. separation

A grounding system protects the user against being electrocuted during ground faults (a short circuit that occurs when a hot wire touches a ground or noncurrent-carrying conductor). It also protects both the user and the appliance against the buildup of induced voltages and provides a safe discharge path for surges.

Grounding is basically a two-part system. The first part consists of a wire present in all the branch circuit electrical cables that connects the frame of all the appliances, tools, and metal outlet boxes back to the grounding bus of a fuse or circuit breaker panel. This wire is called an equipment grounding conductor. It is the bare wire in the NM cable and the green wire within a conduit system. It could also be the armor of an AC cable if the metal conduit runs all the way back to the service panel. The second part is the earth grounding system. It is composed of a large copper wire, called the grounding electrode conductor, and a series of ground rods (or equivalent) that are buried in the earth.

Grounding Wire Connection

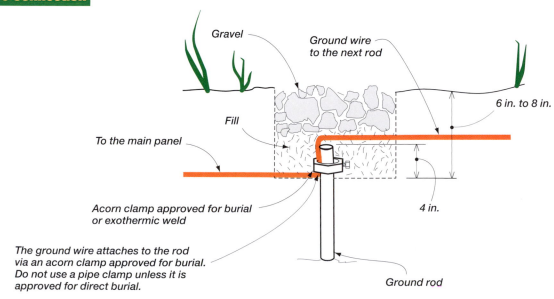

Gravel

Ground wire to the next rod

Fill

6 in. to 8 in.

To the main panel

4 in.

Acorn clamp approved for burial or exothermic weld

The ground wire attaches to the rod via an acorn clamp approved for burial. Do not use a pipe clamp unless it is approved for direct burial.

Ground rod

Methods of Grounding

There are many methods of grounding, such as using structural steel, rebar, or well casings. However, ground rods are used for most residential applications. But sometimes rods cannot give you a low-resistance ground due to low moisture in the earth or excessive rock and little soil. Below are methods of grounding from the worst to the best:

- Code minimum (unless you can prove 25 ohms or less with one rod). Two rods are considered inadequate for today's electronic equipment.

- My Above Code requirement. Eight rods, 8 ft. to 20 ft. apart, depending on the available area. I also tie in the building's metal well casing, if available.

- A ground ring (#2 or larger copper bare cable circling the house). Ground rings work well but aren't that practical. In most residential situations, you cannot open a trench around the building, because it would be in the way of the rest of the building construction. You also risk cutting other utilities if they are already in place.

- A circle of ground wire under deep-driven pillars or posts. If deep support posts are going in, take advantage of the opportunity to obtain a low-resistance ground. Form a long continuous circle about the size of a dinner plate and stick it under the post.

Code Minimum

Panel connection

Ground rods

Above Code

Panel connection

Well casing

Ground Ring

Panel connection

Ground ring

Under Post

Panel connection

Ground wires to support posts

Support post

House

Earth

Support post

Ground wire under support posts

Protection through Grounding

Grounding offers us life safety, equipment safety, and longer equipment life. The equipment-grounding conductor works like this: If a hot wire touches a grounded appliance frame, grounded tool frame, or the side of a grounded metal outlet box, the fault current will ride back to the service panel via the equipment grounding conductor, complete the circuit, and trip the breaker. The ground rod and earth ground play no part in this system.

The earth grounding system comes into play whenever there is any static voltage buildup on an appliance's frame. A typical example would be a lightning strike. The strike puts out a massive electromagnetic force (EMF) pulse that induces a voltage into metal appliance frames. The equipment-grounding conductor takes this surge back to the main panel, where it transfers to the grounding electrode conductor (ground wire), which sends it to the ground rods to bleed harmlessly into the earth. This same large ground wire is also used to transfer surge pulses (caused by the utility company opening and closing a circuit, a tree falling on a line, a car hitting a pole, etc.) to the earth that have been stripped off the line by arrestors mounted at the panel or at the point of use—like your computer's surge protector. (For more information about lightning strike protection, see chapter 13.)

None of what grounding has to offer will make an appliance work any better if these problems aren't occurring. Indeed,

Ground Fault on a Grounded Appliance

Ground Fault on an Ungrounded Appliance

The breaker will remain on.

Hot

Neutral

Main panel

To the ground rods

Ungrounded receptacle

The frayed cord shorts to the washer frame.

Washing machine

Concrete floor

Frame energized

electrical gadgets can work just fine without proper grounding. Perhaps that's the real reason grounding took so long to enter residences: It was hard to convince people that an appliance needed something when it was working just fine without it. Grounding is akin to a personal insurance policy. You never know when you'll need it, but you sure want to have it when you do.

Ground fault

If a hot wire accidentally touches the grounded frame of a tool or appliance and energizes it, a ground fault has occurred. This is one of the more common types of tool/appliance malfunctions. For example, let's assume that the cord of a grounded washing machine has frayed (the washing machine is sitting on a wet basement floor) and the hot wire is touching the metal frame, resulting in a ground fault. Once the

A deadly ground fault. A too-tight NM connector has cut into the cable at the pressure switch (top left of photo). The multimeter reads 120 volts to ground on the water line.

An Ungrounded Can Opener

Some time ago, I was troubleshooting a problem where a customer was getting shocked every time she leaned over the sink to wash the dishes (the sink was nonmetallic). I traced the problem to a metal trim ring around the sink. When I lifted the electric can opener off the ring, the problem went away. Turning the can opener over, I discovered that one of the screws attaching one of the rubber feet to the opener extended beyond the rubber pad. A hot wire inside the opener had shorted to the metal frame on the opener's bottom and, because this was a two-wired (ungrounded) appliance, the short was being transferred through the screw into the sink trim ring. If the ground prong were present, then the short would have returned back to the main panel's neutral/grounding bus and kicked the breaker. Note that the earth is not in the current flow path during a ground fault.

frame is energized, the fault current will flow back to the main panel via the equipment grounding conductor and the circuit breaker will trip off. In this situation, the fault current does not flow to earth ground to dissipate. It remains on the low-resistance wiring to the breaker, allowing the breaker to trip off.

If an equipment grounding conductor is not present, if a cheater plug is used (I see this all the time), or if there is a bad connection somewhere along the equipment grounding conductor, the breaker cannot trip because there is no way for the fault current to get back to the main panel to complete the circuit. The frame of the washing machine will remain energized, and a person who touches it can receive a serious shock. What's happening here is that the person's body becomes an electrical conductor to complete the grounding conductor path.

A ground fault doesn't have to be direct, either. Though the initial fault or short to the appliance may be direct, the routing back to the panel may be of high enough resistance that it lowers the

current below the level needed to kick the breaker off. Although the current may not be high enough to trip a breaker, it could still be lethal to both you and the appliance.

Unwanted voltage and current

If an appliance frame isn't grounded, the 0-volt potential (on its frame) that we need for safety cannot be maintained. Without grounding conductors to direct the unwanted current away and an earth grounding system to dissipate it, the stray voltage can build up to a damaging potential—arcing over to us if we touch it or through the appliance itself.

Stray current resulting from induced voltage can come from any current-carrying source (for example, current through one wire can induce current into another). Induced voltage also comes from any continuous voltage source, such as an overhead high-voltage line. When a good grounding system does not exist, a power line can induce voltage and direct current into a home with a metal roof or metal framing. I've seen this problem in mobile homes that are close to high-voltage transmission lines. The siding's metal skin can build up to some very high voltages and give you quite a shock. To avoid this type of voltage buildup, the metal must be well grounded into a system of ground rods.

Surge current comes primarily from lightning. The magnetic field created by a lightning strike, even if it's over a mile away, can induce voltage into any current-carrying conductor (such as house wiring) or noncurrent-carrying conductor (for instance, metal water pipes and appliance frames), producing a short-lived high-voltage spike. If the induced voltage and current find no designed-in path from the appliance to ground, it will find its own, such as through sensitive electronic equipment, or even arc through the air. The resulting shock to humans can vary from a minor jolt to electrocution.

Surge arresters

Surge arresters suppress voltage surges (that's why they're also called suppressors) by directing excess voltages and currents into the grounding system and thus into the earth. But neither a high-quality surge arrester in the main panel nor a point-of-use arrester at the appliance will work properly without a high-quality grounding system to dissipate the surges. Without that, the surges simply go right into the equipment you are trying to protect. And the odds of it picking that new large-screen TV rather than an old black-and-white set are almost 100%.

Choosing the Proper Grounding Materials

Now that we know what a grounding system is and what it can do for us, the question is how to build one. Choosing the right materials can sometimes make the difference between a properly working grounding system and one that doesn't work at all.

Ground rods

Ground rods connect the grounding system to the earth. Manufacturers make many different types of grounding rods, but the two most common choices are galvanized steel and copper clad. Both types will do the job.

DIAMETER Code requires a minimum of a ⅝-in.-diameter rod. The next step up is ¾ in. and 1 in. The latter two are only available at large electrical wholesalers. In

general, I don't think the larger-diameter rods are worth the extra expense.

Avoid the little 4-ft.-long, ⅜-in.-diameter rods that are sold at electronics stores and discount houses. These minirods aren't legal for anything. I've noticed that people use these rods to ground individual items, such as a TV antenna. Don't do it. The antenna, the telephone system, the cable TV system, and everything else must be tied into the main house grounding system at the service panel.

LENGTH The rods must be at least 8 ft. long; if you can find them, they do come in 10-ft. lengths. Special clamps are also available; they allow you to stack ground rods on top of each other. Deeper is better, but it is debatable whether or not the extra length is worth the extra expense and effort. In most cases, the extra depth does not significantly lower the ground resistance, unless it gets you into a moisture bed. It's better to drive a second rod (NEC requires a minimum spacing of 6 ft.).

In cold areas of the country with deep frost lines, it may pay to go deeper. Low temperature yields increased ground resistance within the frost-line area. Ground rods must be long enough to go deeper than the frost line. In that case, extra-long rods may have an application. If possible, consider coupling ground rods together or taking advantage of a metal well casing.

Other grounding options

At one time it was common practice to use a metal plumbing system in lieu of ground rods. Many homes are still grounded that way. If this is the case with your home, change it immediately. Code does allow metal plumbing lines to be used as a ground rod if three conditions are met: a ground rod system is also pres-

■ WORK SAFE
■ WORK SMART
■ THINKING AHEAD

It's imperative that you bury the entire ground rod. Leaving one end of the rod sticking out of the ground is a trip and lawnmower hazard.

ent; the pipe is in contact with at least 10 ft. of earth (which is impossible to know); and the connection to the pipe is within 5 ft. of where the pipe enters the building. However, should a plumber replace the metal pipe with a plastic one, he'll nullify the ground. To be safe, I always think of the metal plumbing system as something that needs to be grounded, not something to ground to.

Rebar is also sometimes used instead of a ground rod. The footer, connected into the grounding system via rebar, is an effective low-resistant ground. To use it, the rebar must not have a nonconductive coating (in my section of Virginia, nonconductive coating looks like light-green paint) and must be in contact with at least 20 ft. of concrete. In addition, the concrete footer must be in direct contact with the earth, as opposed to having a plastic barrier underneath it.

Another problem with using rebar is connecting it to the ground wire. An acorn clamp connection, which simply presses the copper wire against the metal of the rebar, rusts quite quickly and doesn't provide a good long-term ground. If you want to use the rebar as a grounding electrode, you should use an exothermic weld system; unfortunately, these welding devices can be difficult to find and tricky to use. Since it's not possible to verify that all of the above conditions are met, and because of the problems associated with the rebar/clamp attachment, I rarely use rebar in lieu of a ground rod system.

Ground wire (grounding electrode conductor)

Because aluminum corrodes so easily, I use only copper ground wire. Although the ground wire from the service panel to a ground rod never has to exceed #6 copper, I use #4 copper most of the time, because it has a significantly larger diameter and thus has less resistance, especially on long runs.

For 200-amp service, you need #4 copper to connect to the metal plumbing, rebar (if used as a grounding electrode), or anything else you may want to bond into the grounding system. For a typical eight-ground rod system, you need around 60 ft. of wire, assuming you don't have to make a long run to get into that moist (low ground resistance) valley.

Clamps

Make sure that the clamp you use to connect the ground wire to the ground rod is approved by UL for direct burial. The common clamp used is called an acorn clamp. Do not use hose clamps or the toothed clamps commonly used to attach wire to pipes inside the house—they are not approved for direct burial and quickly rust and corrode.

I once measured the clamp resistance of a large number of grounds for a utility cooperative that was trying out some new equipment to measure clamp- and ground-resistance. The average resistance through their corroded toothed clamps was 140 ohms (that's a lot of resistance). A several-thousand-amp surge (a typical lightning pulse) through such a high-resistance connection can develop several thousand volts. This high voltage is felt throughout the grounding system, and it will arc *through* sensitive electronic equipment if it finds a lower resistance path to ground. For some unknown reason, this normally occurs through the most expensive electronic gear, such as a color TV. With the approved clamps, you can expect a ground clamp resistance of close to 0 if it is installed properly.

You are allowed to skip using the clamp system entirely and instead use an exothermic weld system to connect the ground wire to the rod. Such a system

Acorn clamps are the most common clamp used and are approved for direct burial. Note that the ground wire is in the back of the clamp—in the V—not under the screw.

Brass toothed clamps. Use these to grab onto the water lines inside the house to ground them, not onto the ground rods outside.

actually welds the wire to the rod, so you never have to worry about a bolt coming loose on the clamp. However, I no longer use this system due to problems with the wire snapping off the rod (see Attaching the Clamps on p. 106).

Installing a Low-Resistance Panel-to-Earth Grounding System

Installing a low-resistance panel-to-earth ground to instantly drain away unwanted current flow and keep the ground system at 0 volts is not an easy task. Absolutely everything is working against you in obtaining a low resistance ground: rocky soil, low moisture in the earth, poor conductivity in the soil, human error, and so on. Thus, your objective in designing and installing a grounding system is to overcome these problems to the point where

life and equipment are no longer in jeopardy. In common residential construction, a good grounding system starts with the actual earth connection itself—the ground rods.

ABOVE CODE Ground rods (also called grounding electrodes) are normally driven into the earth close to the main panel or utility meter base. Code demands two rods unless you can prove a resistance of 25 ohms or less (many homes only have one rod, which is not enough in most cases). I install eight rods because I'm always after the lowest ground resistance I can get. In theory, doubling the number of rods should reduce the ground resistance by half. In reality, it varies and can be significantly more. Regardless, the bottom line is the more rods the better. Since ground rods are cheap, normally less than $10 each, there is no reason to obtain anything other than the best ground possible.

Rods must be in contact with moist earth for their full length to get a good low-resistance ground. This ground rod was installed in the rock fill around the footer because it was easier to drive down. The homeowner will have a higher ground resistance as compared to a rod that is completely surrounded by soil.

Installing ground rods

Once you have all the materials, the question is where to drive the first rod. Though the first one or two rods should be driven reasonably close to the main panel to appease the inspector, the rest should be driven where the ground is moist and can provide a low ground resistance. A good location close to the house is normally around the gutter downspouts and the drip edge. To obtain a lower resistance ground, consider running the ground wire to a creek bed, dry pond, or wooded valley. When driving rods into a creek or pond bed, do not use an AC-powered hammer drill (for obvious reasons). Use a sledgehammer or a battery-powered hammer drill.

Don't cheat when you install the ground rods. What I mean by cheating is driving the first rod into the loose fill adjacent to the footer or support post—so loose that you can easily shove the rod into the ground bare-handed. If the rod

Driving Ground Rods in Rocky Areas

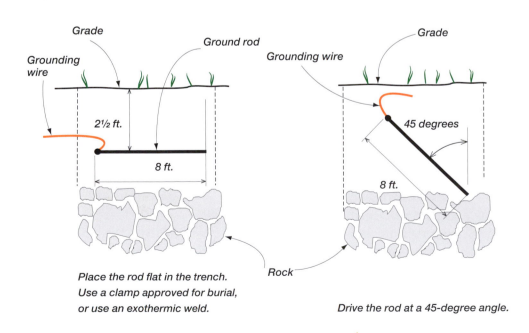

Place the rod flat in the trench. Use a clamp approved for burial, or use an exothermic weld.

Drive the rod at a 45-degree angle.

Dig a trench, drive the rod at a 45-degree angle, and then bend the rod in the trench.

goes in easily, odds are you do not have good, tight contact with the earth and will wind up with a higher ground-to-earth resistance.

In rocky areas, you may be lucky to get down 2 ft. If you can pull the rod back out when you hit a rock, do so. If not, cut it off just under the final grade and use it as a ground, but keep driving more rods. The more rocks you hit, the more rods you will have to install. If you have no significant depth at all, consider laying the rods in a trench and bringing in dirt to cover them. Covering the rods with rock will do you no good—you need soil.

For rocky areas, I highly recommend using a rotary hammer to drive the rods. In normal soil conditions, it can drive an 8-ft. rod in less than a minute. In rocky areas, the many hammer blows it produces can split a small- or medium-sized rock. If you hit a large boulder, you're still out of luck.

If you're unable to drive the rod all the way in because you encounter solid rock, you've got three options other than just cutting the rod. You can drive the rod at an angle up to 45 degrees from vertical or lay the rod in a trench (minimum 2½ ft. below the surface, or grade). Another option is to dig a trench and drive the rods at 45 degrees until you hit rock. At that point, bend the rod flat into the trench.

The distance between the rods is very important. Code requires a minimum of 6 ft., but I feel this is not enough. You want the distance between the rods to be at least equal to the length of the rod (normally 8 ft.) to as much as twice that distance. What you are trying to avoid is the "sphere of influence," where the electrical field around one rod affects the electrical field around another. This increases the resistance of the electrical current flowing through the soil.

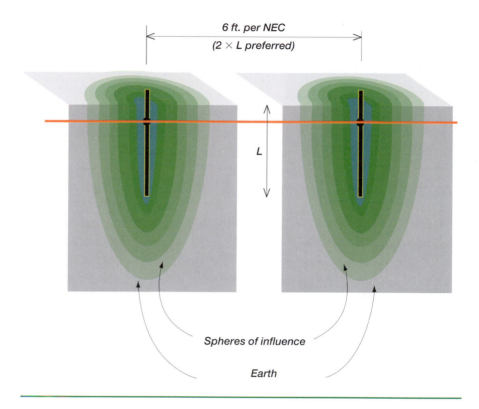

Grounding

6 ft. per NEC
(2 × L preferred)

L

Spheres of influence

Earth

Once you've found a suitable location, use a post-hole digger to dig a hole 6 in. to 8 in. deep for each rod. Drive the rod, leaving about 4 in. above the bottom of the hole. As I mentioned earlier, the easiest way to drive a ground rod is with a hammer drill (you can rent one). The most common way to drive a rod is with a sledgehammer. The problem with using this method, however, is that the rod swings as you try to hit it. To help hold the rod steady, drill a 1-in. hole through scrap board and put it over the rod or make a hook using PVC pipe. Try not to mushroom the end of the rod, or you'll have to file it to fit the clamp over it.

Running the ground wire

Once all the ground rods are driven, open a ditch between them with either a pick or a mattock. Lay the ground wire in the

■ WORK SAFE
■ WORK SMART
■ THINKING AHEAD

Smooth all the sharp bends out of the ground wire as you unroll it along the ground. This allows the acorn clamps to slide easily along the conductor.

■ WORK SAFE
■ WORK SMART
■ THINKING AHEAD

For a better ground connection, use two acorn clamps and apply thread lock compound on the bolts to keep them from working loose.

■ WORK SAFE
■ WORK SMART
■ THINKING AHEAD

Do not use electrician's pliers to tighten the ground clamp—they can't get it tight enough. Use an end wrench, an adjustable wrench, or a socket wrench.

ditch and cut it to length. Cut the wire about 1 ft. longer than what it lays out to be, just to be on the safe side. Do not cut the wire anywhere along the run—it must be continuous from beginning to end, regardless of the number of ground rods you plan to install. (This is why I buy the wire in rolls and then unwind the length I want.) Slip the acorn clamps onto the wire. Slide them along the wire to the approximate location of the ground rod, then slip the clamp onto the rod as you push the wire into the ditch.

For protection, the wire must also be buried (at least 6 in. deep) along its run and enter the wall cavity as soon as possible. If the wire along the outside of the residence is not subject to damage, it can be run without conduit as long as it is stapled flat against the wall approximately every 2 ft. (the wire can be painted to match the building's color scheme); if the wire can be easily damaged, run it through conduit.

Attaching the clamps

The clamps need to be properly fastened to the rods. To do this, run the wire parallel to the rod for about 1 in. before and after the clamp. Do not run the wire into

and out of the clamp from the side; doing so will result in a loose, high-resistance connection. If you are using an exothermic weld system, be careful not to bend the wire close to the weld. The heat produced by the welding process makes the wire brittle and easy to snap, which is why I no longer recommend this system.

Cover the ditch once the clamps are attached. As you fill it in, push the wire down to ensure that it's at the bottom. If aesthetics allow, fill the ground-rod hole with gravel to allow the rain to flow through the gravel and saturate the soil around the rod for an even better ground. If you located the hole close to a gutter, then the gutter water will saturate the earth around the rod. (See the top drawing on p. 96.)

Bonding the panel frame

The main panel frame, being metal, will always have a panel bond, which is simply a wire or metal strap that connects the panel into the grounding system. An inspector normally makes a point to verify its presence. Some panels use a screw that is turned down through the grounding

If you use a sledgehammer to drive the rods, you'll need to steady the rod as it is hit. Never use your hands. Instead, have someone steady it with a board that has a hole drilled in it or a hook made out of plastic pipe.

This ground wire is entering and leaving the clamp from the side (as opposed to dead center down the V back). In a few weeks or months, the ground wire may slide to the center of the V, creating a loose, high-resistance connection.

A missing ground wire. The wire that used to run to another rod has snapped off the right side of this one. The wire was made brittle by the exothermic weld, and when bent, easily broke free of the connection. When this happens, you must cut off the top of the rod, as I did here, and start from scratch.

bus and into the panel's steel; others have a bonding wire that runs from the bus through a prepunched hole in the steel frame. Although the inspector may only check the bond in the main panel, all metal box frames in the electrical system need to be bonded into the grounding system, including the cutoff boxes for heat pumps, water heaters, and submersible pumps.

What Needs Grounding

When it comes to grounding, remember this rule: If it can become accidentally energized—through an induction or a fault—ground it. I've been shocked by sheet-metal ductwork, water lines, drain lines, and metal siding—if it's metal, it's bitten me. Therefore, I require all ductwork to be grounded. Even if a heat

pump or furnace frame is grounded by its own equipment-grounding wire, the ductwork may not be. Many times there is a nonconductive flex joint between the unit and the ductwork that acts as an insulator. I also recommend grounding the rails of garage-door openers to prevent lightning surges from building up and destroying the electronic opener. Following is a list of things that I recommend be grounded.

Metal water pipes

To ground metal water pipes, make the primary connection from the pipe to the grounding bus within the main panel and connect it to the pipe on the house side of the main valve. You'll also need to run bonding jumpers where needed.

Bonding Jumpers

The bonding jumper should be run around the water heater because some heaters are made of nonconductive material, which will interrupt the ground circuit. The jumper also allows the heater to be replaced without disrupting the ground connection.

If your water pipes are metal, including the feeder from the city connection, run the jumper around the meter.

The bonding jumper is a separate section of wire with clamps on each end; it allows the grounding conductor to be run around anything that could interrupt the grounding circuit. For example, you need to run a jumper around a water heater because some water heaters are now made of nonconductive material. Also, water heaters get replaced, and you don't want to disconnect the ground at any time. If a jumper is not there and if fault current is flowing in the water pipes (this is quite common), you can be electrocuted when you change the water heater. If all pipes are metal, including the feeder from the city connection, install a jumper around the meter.

Metal conduit

All metal conduit within the house must be grounded. I normally don't depend on conduit to carry the ground, because there is too much of a chance that a slip-on fitting will loosen and open the grounding system (screw-in fittings make better grounds). I normally run a grounding wire within the conduit as much as is practical. Consider using plastic conduit if you don't want to worry about grounding it.

In a residence, metal conduit is most commonly used in the garage and basement. Short sections of thin-wall conduit protect the wire as it runs from the receptacle and switch boxes vertically up to the floor-joist system. To ensure a proper ground, connect the metal conduit tightly to a metal receptacle box and, in turn, ground the box via the grounding wire in the NM cable entering the box. However, the particular knockout hole you want to use must have a single knockout hole as opposed to multiple ones. The punched metal surrounding a multiple knockout is not a reliable conductor of fault current (thus the use of a grounding bushing and a jumper to bridge the gap).

Receptacles and switches

Receptacles and switches in nonmetallic boxes can be grounded in several ways. The standard way is to simply connect the incoming bare grounding wire onto the outlet grounding screw. If there are two or more cables, cut the ground wires from the cables to the same length and add a 6-in. bare wire to the bunch; this will serve as a jumper from the splice to the receptacle. Using broad-nosed electrician's pliers, twist the ends of the wires together in a clockwise direction for about 1½ in. to 2 in., then twist on a wire connector. The wire connector does not have to be green.

Some inspectors require crimped ground wires in the outlet boxes (the NEC does not). Although their intent is good (to provide the best ground connection), the end result is not. Your chances of having a loose ground splice due to a bad crimp are greater than having a loose ground due to a bad wire-connector splice. Personally, I don't crimp my grounds because I want to be able to add or remove one later if I change the wiring design. You can't do that with a crimp system—you have to cut every ground wire at the crimp. In my opinion, if wire connectors are good enough to splice hot conductors, they are certainly good enough to splice ground conductors. If a crimp system is used, you must use

Grounding Metal Conduit

Thin-wall metal conduit

Wall

Make sure the screw is tight to the ground vertical conduit.

Metal receptacle box

NM cable

Self-threading machine screw is painted green to indicate that it's a ground connection, not a means of attachment.

An equipment grounding wire grounds the box and conduit connected to the box.

Attachment screw

This box has prepunched holes for both single and multiple knockouts. A single knockout hole can be used to ground conduit via a grounded box. Multiple knockouts need a grounding bushing, because fault current may not get through the area of the punched metal.

Grounding Receptacles in Nonmetallic Boxes

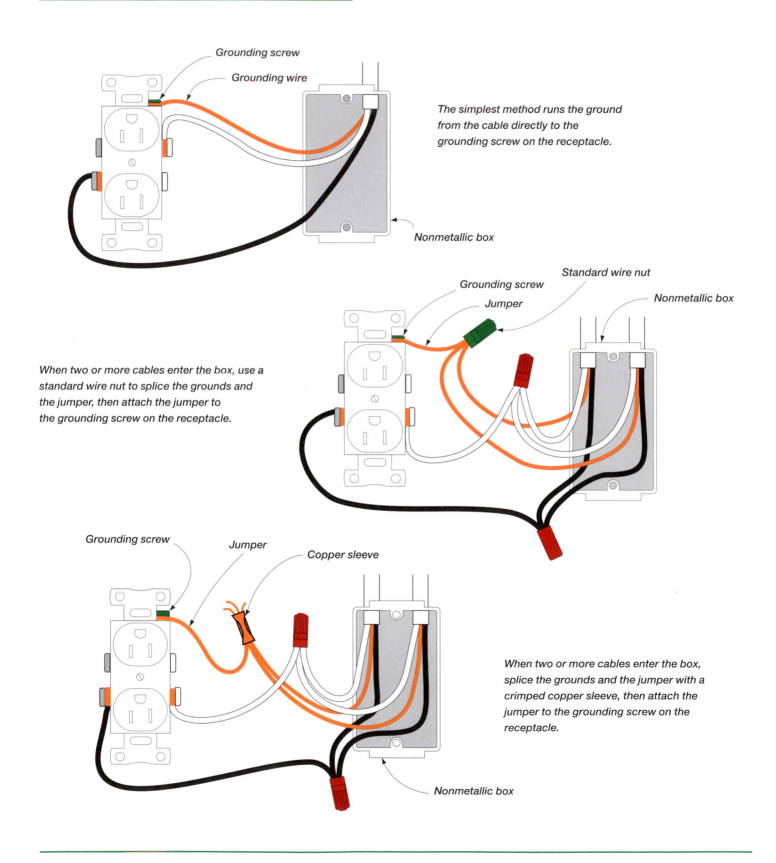

Grounding screw

Grounding wire

The simplest method runs the ground from the cable directly to the grounding screw on the receptacle.

Nonmetallic box

Standard wire nut

Grounding screw

Jumper

Nonmetallic box

When two or more cables enter the box, use a standard wire nut to splice the grounds and the jumper, then attach the jumper to the grounding screw on the receptacle.

Grounding screw

Jumper

Copper sleeve

When two or more cables enter the box, splice the grounds and the jumper with a crimped copper sleeve, then attach the jumper to the grounding screw on the receptacle.

Nonmetallic box

A four-way crimp system. I don't recommend crimping ground wires, but some inspectors require it. If you must crimp, be sure to use the proper tool; don't use pliers or a common crimper, which just mashes the wires together in one place.

the proper crimping tool: a four-way or four-point crimper.

Another method utilizes a green grounding wire connector that has a hole in its end. Forget using the hole—it's more trouble than it's worth. If you want to use a green wire connector like a typical wire connector on the splice, go ahead. However, the color green is not required for the splice.

You can ground receptacles in metal boxes via the grounded metal of the box in lieu of installing a wire jumper to the ground screw. To do this, you must have direct metal-to-metal contact between the box and the metal ears of the receptacle. Some receptacles come with special clips on one ear just for this purpose. This grounding rule does not apply to metal boxes where the receptacles attach to a removable cover and not to the box.

Metal boxes

A metal box must always be grounded. The ground can come via a piece of grounded conduit if the box knockout is a singular hole (you cannot obtain a

Double-Checking the System

Grounding connections are like Christmas-tree lights—one problem, and nothing works. If you had a ground-resistance-testing machine, called a megger, you could check the overall system, but most of us don't have one. Here is a checklist to help you to verify that everything is in working order:

- Don't just wrap the ground wire around the rod or use hose clamps. Use an acorn clamp and install it properly.
- Leave the first hole open so the inspector can observe the clamp connection.
- If you can't get the ground wire inside the wall immediately, be sure to clamp it tight against the siding.
- Use copper ground wire.
- Don't use gauge less than #6 to the rod and use #4 copper for everything else. Make sure all connections are tight.
- Do not place the ground rod(s) under a paved driveway.
- Bond the metal gas and water lines to the panel with #4 copper.
- Bond the panel's grounding bus to the metal frame of the service panel.
- Attempt to place as many rods as possible in a moist area.
- Use a grounded bushing on metal conduit connecting the service panel to the meter base.
- Add a metal well casing, if there is one, into the grounding system.

If you're thinking of plugging in a grounded appliance to an ungrounded receptacle using a cheater plug, don't do it. The cheater plug is simply an adapter that allows you to insert a three-prong plug into a two-prong outlet. Cheater plugs cannot ground an appliance unless the receptacle box is metal and there is some type of conduit providing a ground all the way back to the main panel. (This is rare in residential situations, but it does occur in older houses wired entirely with BX or Greenfield cable.) If the appliance does not have a proper ground connection and a ground fault occurs, someone could be electrocuted.

To have yoke-to-box ground contact, you must remove the insulator bushing. This one is orange. However, code now limits such use to surface boxes only. Thus, this is a violation for this box.

High-quality receptacles have a built-in spring-type grounding strap that automatically grounds the receptacle when the attachment screw enters the grounded metal box.

ground through a multiple knockout) or via an equipment-grounding conductor. In a residence, you normally use the latter. The conductor terminates in a splice where a jumper leaves to ground the box via a screw. These jumpers are called pigtails; they can be purchased with screws already attached if you don't want to make your own.

There are a few things you cannot do. First, you cannot attach the ground pigtail to the box via the screw that secures the box to a stud or a wall (it's against code); in this case, use a separate screw. Second, you cannot run the jumper to

the receptacle rather than to the box, and then think that the box is grounded through the receptacle. And you cannot use a sheet-metal screw to ground the box (it doesn't have enough threads).

Appliances

Until 1996, the NEC allowed some appliances, in particular electric dryers and stoves, to use the neutral as the ground as well, and you could get away with using a three-prong plug. Now, all stoves and dryers must use an equipment-grounding wire that is separate from the insulated current-carrying neutral.

Many smaller appliances are grounded via a three-prong plug and cord (the roundish prong is the ground connection). However, for an appliance to be truly grounded via the plug and cord, a grounded three-prong receptacle must be installed with an equipment-grounding conductor connecting it back to the main service panel grounding bus. To make sure the receptacle is grounded, check it with a plug-in tester.

The biggest problems come when you have new appliances and old wiring. If you buy a new appliance that has a grounded plug (or a hard-wired appliance that has a green wire in its splice box), it requires grounding via the manufacturer's instructions; otherwise, you have the potential of electrocution. The most common violations occur with common appliances, such as dishwashers and water heaters. There is no grandfather exemption. If you have an older house with ungrounded wiring, you must either run a ground wire back to the panel or run new, grounded cable.

Submersible pumps

If you've got a well, its submersible pump and metal casing must be grounded. The metal well casing, around 6 in. in diameter and driven through soft soil and into bedrock, is arguably the best ground medium available. That is, it has the lowest resistance to ground of anything around. The NEC requires that you simply attach the equipment-grounding wire from the pump to the metal casing. In high-lightning areas, it's best to supplement the code-mandated ground connection by running a 4- or 6-gauge copper wire from the casing to the ground-rod system. I also install a Tytewadd surge arrester at the ground attachment to provide surge protection for the pump. (See chapter 13 for more on lightning suppression.)

Structural steel

Exposed structural steel, such as a building frame, should be bonded to the grounding system. The metal siding, if any, receives a ground through its contact to the beams. The bonding jumper needs to be accessible where the jumper is connected to the frame. Although you can connect the bonding jumper to the ground rods or a grounding electrode conductor, I prefer to connect it directly to the main-panel grounding system so I don't have to worry about any bolted connections coming loose. The conductor is sized according to tables within the NEC (for 200-amp service, use 4-gauge copper). Although it's not required, metal siding and metal roofs on buildings with no structural steel can be grounded to the main panel.

Gas pipes

The NEC forbids the use of a gas pipe as a grounding electrode. Whether or not gas pipes need to be grounded depends on the local code; however, the NEC requires metal gas lines to be bonded to the house grounding system. I always ground the gas pipes because it makes good sense. Simply run a #4 copper wire to the metal pipe on the house side of the turn-off valve. You can use the equipment-grounding conductor of the cable feeding the appliance. If you ground the furnace, for example, and the pipe is screwed into the furnace, it is automatically grounded and you don't need to run a separate ground wire. Be aware that this equipment-ground wire is much smaller in gauge, as compared to a #4 copper ground wire, so you'll be safer with the #4 in addition to the equipment-grounding conductor.

Grounding Multiple Panels

Acceptable

Preferred

Two 200-amp service panels

Two 200-amp service panels

Grounding electrode conductor

To other ground rods

To other ground rods

Multiple panels

It's common knowledge that a service panel must be connected to a ground rod or an equivalent system. But what do you do with multiple panels, which you may have in a 300- or 400-amp system? The common method is to run a grounding electrode conductor from one panel to the next and have one panel connected to a ground rod system. The problem with this logic is that the noise and surges that are stripped off the line in one panel are sent to the next on that same wire. Normally, there may not be a problem, but if you have some sensitive electronic gear in the second panel, there may be trouble. The noise and surges may not destroy equipment in the second panel, but they could overstress them and shorten their life expectancy. And if a big surge—a several-thousand-volt blast stripped off the line by a surge protector in a nearby panel—does come through, the last thing you want to do is send it to another panel. You want it to go straight to earth. The only way to do that is with a direct connection. Therefore, the best way to ground multiple panels is with independent ground wire systems. That is, each panel has its own grounding electrode conductor connected to the same ground rod system.

CHAPTER 6

Wiring Room by Room

ONCE YOU'VE installed the service-entrance system, passed your inspection, and gotten power from the utility, it's time to install the house wiring system. In other words, now that you've run the really big cable, it's time to run the little ones. If you know what you're doing, you can do the job yourself in about a week. The work will go easier and faster if you can enlist a helper to drill holes in the studs and pull cable through the studs and joists.

Planning is key. First, know what materials will be needed and have them on site. This also includes the light fixtures. Store all materials in the center of an unused room. Storing them in the center allows access to the stud walls. Second, locate all outlet locations and determine how you're going to get to them. Third, pull the cable. Through trial and error, all electricians develop tricks and methods to get them through the problems of getting the wire there; I've included as many of these techniques in the chapter as I could.

Stocking Up

A typical 2000-sq.-ft. house uses approximately ten 250-ft. rolls of 12-2 gauge wire: two rolls of 14-2 and two rolls of 14-3. Try to avoid using 12-3 because it is very expensive. I use 12-gauge for all the receptacle circuits even if code allows 14. The 14-3 will be used for interconnecting the smoke alarms and for wiring the three- and four-way switching circuits.

You'll also need UF cable if you are running any outside circuits. If the house has an electric dryer and a water heater, you'll also need 10-3 for the dryer and 10-2 for the water heater. You can use leftover 10-3 for the water heater; just cut the red wire dead (cut it back to where the sheathing has been trimmed). Lastly, you'll need 6-3 copper for an electric stove.

ABOVE CODE Once the cable hits the job site, you'll find that it's easy to get the different gauges mixed up—especially 10, 12, and 14. To be able to tell at a glance which gauge is which, color-code the cable and the box it comes in. This system is especially helpful once installed, especially for the inspector. For example, green could mean 10 gauge, 12 could stay white, and gold could be 14. (You won't confuse any other cables because the physical size is different.) Remove the cable from the box, lay the still tightly rolled cable down on the subfloor or hang it up, and spray both sides. Once the cable has dried, put it back in the box and paint the box.

Design Considerations and Outlet Box Locations

Overall, code requires at least one receptacle within 6 ft. of a door and then every 12 ft. thereafter. The fixed-glass section of a slider door and fireplaces should be treated as a solid wall. Any wall 2 ft. or wider requires an outlet.

Code allows certain details that I try to avoid whenever possible. For example,

code allows standard receptacles to be placed in the floor face up, but you shouldn't do that unless you have to. Floor-mounted receptacles are a catch-all for liquids (from spills or cleaners), trash, and dust. This buildup will destroy the receptacle or may start a fire. If you must have a floor receptacle, invest in a commercial unit that has a cover or a tombstone design (it looks like a tombstone rising out of the floor), which keeps the receptacle vertical.

ABOVE CODE   Code also allows switched receptacles in lieu of switched lighting in some rooms. Ignore that. For room lighting, I always prefer switched overhead. True, switched receptacles save the material and labor costs of installing an overhead light, but doing so commits the homeowner to using one light on that one end table forever. Overhead lights not only provide better overall illumination, they also allow you to move the furniture around.

Color-coding the different cable gauges prevents a gauge mixup during construction.

Code Minimum Placement of Receptacles along Wall Sections

The fixed pane of a slider is treated as a wall.

Window

Door

Sliding glass door

A slider pane is treated as a door.

A receptacle must be placed within 6 ft. of a door frame.

6 ft.   12 ft.   6 ft.   6 ft.   12 ft.

Planning outlet and switch locations

Building a house is a stressful time for most couples. One way to prevent blowups is to have both partners lay out the locations of receptacles and switch boxes together. Otherwise, if one partner picks the receptacle and switch locations alone, he or she should be prepared to hear, "Why didn't you put one here?" or "What about my lighted curio cabinet?" To that, the only answer is, "Whoops."

Once the framing is done, I advise couples to walk through their home with a big black marker and put an X on the stud where they want the outlets. If you want an outlet on a particular side of a stud, put an X on that side. Don't worry about the height of the outlet box just yet. Take your time—bring a couple of folding chairs so you can sit and talk about where the furniture is going and where the outlets will best be placed.

Don't forget to mark the switches. To do this, you must know which way the door is going to swing. Sometimes it will be obvious, as when the door

opens against a wall, but when the door opening is dead center in the wall, check the blueprints and get a verbal confirmation from the head carpenter, then mark the direction on the floor to avoid changes later.

The light switch should be on the wall adjacent to the latch side of the door. Never mount the switch outlet box too close to the edge of the frame; otherwise, it may get in the way of the trim. To avoid this, put the box on the next stud back from the door-frame stud, or scab on one or two pieces of scrap wood to the door-frame and mount the box to that.

Now consider switch and outlet heights. Switch outlet boxes should be 48½ in. to 48⅝ in. from the top of the upper plate. Assuming that the drywall is going to be installed from the ceiling down, placing the outlet box at this height allows for easier drywall installation. The drywall installers will only need to cut the bottom sheet. Unless you have a specific reason not to, place common receptacle boxes one hammer-length up from the floor. There is no code spec on

TAKE NOTE

As you start marking the receptacle box locations, you may begin to wonder how many outlets you can have on one circuit. In commercial work, we are limited to 10 outlets on a 15-amp circuit and 13 on a 20-amp circuit. However, there are no such limits in residential work—the number is infinite. Common sense, though, tells us to stay close to these numbers or to simply put the receptacles of different rooms together.

Mark the proposed door swing on the floor and have the head carpenter approve it. This eliminates any responsibility on your part if the door is mounted in the opposite direction.

Using a hammer as a height gauge keeps all receptacle boxes at the same level and allows for quick installation.

Code Minimum Placement of Kitchen Receptacles

Having receptacles this far apart normally doesn't provide enough outlets for appliances.

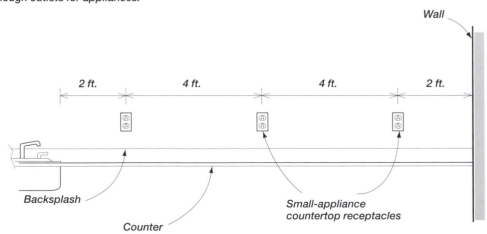

Wall

2 ft.   4 ft.   4 ft.   2 ft.

Backsplash

Counter

Small-appliance countertop receptacles

height here, and using the hammer as a height reference makes everything uniform and speeds things up—just make sure you always use the same hammer.

Kitchen

There are perhaps more receptacles in the kitchen than in any other place in the house—thus, there are more rules that apply here than in any other place in the house. First, I'll discuss the NEC method, or the minimum code approach. Next, I'll go into my Above Code method, which I feel is a more practical approach, considering today's abundant kitchen appliances.

THE NEC KITCHEN Code requires two 20-amp circuits to feed kitchen receptacles. How these two circuits are distributed around the room is immaterial, as long as both circuits are on the countertop. Any receptacle that feeds the countertop must have GFCI protection.

Along the counters, the receptacles must be spaced so that no one spot is more than 2 ft. from an outlet. Practically speaking, this means that when starting from a counter edge, stove edge, or sink lip, you must have an outlet within 2 ft. and then an outlet every 4 ft. thereafter. Also, outlets cannot be mounted faceup in the countertop because water will spill into them. If receptacles are placed above the counter, keep them within 20 in. of the counter. If receptacles are in an appliance parking garage, they do not count as the required number of receptacles.

Kitchen receptacles must be kept unto themselves. For example, loads such as an outside receptacle outlet, in-room lights, or fixed-in-place appliances (for instance, a dishwasher) cannot tap into the receptacle circuits. However, there are a few exceptions. A receptacle installed solely for an electric clock and receptacles installed to provide power for supple-

Optional Horizontal Receptacle Placement

Optional horizontal receptacle location

Optional horizontal nailer

Top of the proposed counter

Wall edge

mental equipment and lighting on gas-fired ranges, ovens, or counter-mounted cooking units are allowed.

Islands and peninsulas must have at least one receptacle outlet if they measure 1 ft. by 2 ft. or larger (if you don't want to install a receptacle, make the measurement 1 in. shorter in either direction). You are allowed to put the receptacle outlet in the counter base but not more than 12 in. below the countertop, as long as the countertop lip does not extend more than 6 in.

Fixed-in-place appliance loads include items such as electric stoves, dishwashers, and garbage disposals. Use #6 copper (50-amp breaker) for the stove. The dishwasher is normally a 12-gauge dedicated circuit. Garbage disposals can be run on either 14 or 12 gauge, depending on what the manufacturer requests. The refrigerator is also considered fixed in place and will need its own receptacle. In this case, the refrigerator can be tied into the kitchen circuit or can have a separate 14-gauge cable. The lights (switched lighting is required) can be tied into any other general-purpose receptacle circuit in the house, such as the bedroom or living room, or put on a separate lighting circuit.

Code-Allowed GFCI Placement

Bathroom #1

Bathroom #2

More baths

GFCI

Above Code GFCI Placement

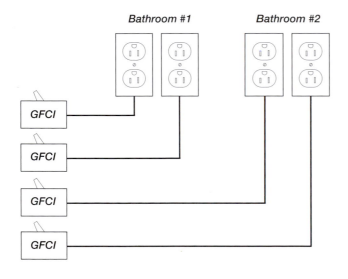

Bathroom #1

Bathroom #2

GFCI

GFCI

GFCI

GFCI

The common along-the-wall outlets must follow the same 6-ft. rule that applies to the rest of the house: Nowhere along the wall can you be more than 6 ft. away from an outlet. This means there must be an outlet within 6 ft. of a door and every 12 ft. thereafter. The fixed glass of a slider is considered a wall.

THE ABOVE CODE KITCHEN The NEC method is good, but I don't think it adequately reflects the trend of today's homeowners, who are using more heavy-wattage appliances. To see this in action, all one has to do is attend a bridal shower and note all the kitchen appliances. And as most of us tend to forget, the NEC requires minimum code and we are allowed to exceed it. Here is my version for the kitchen.

First, unload the countertop appliance circuits. Take the two 20-amp, code-mandated circuits and isolate them to the countertop. Next, bring in an additional circuit for the common wall receptacles. Thus, the kitchen will have a total of three 20-amp receptacle circuits: two GFCI circuits for the countertop (I use panel-mounted GFCIs because of their longer life expectancy) and one non-GFCI circuit for the common wall receptacles. The refrigerator can be tied into the 20-amp wall receptacles or can have its own 20-amp circuit. Along the backsplash of the countertop, place a receptacle on each stud or every other stud. With 16-in.-on-center stud spacing, this places a receptacle every 16 in. to 32 in. If the wall has 24-in.-on-center studs, place a receptacle on each stud.

If you don't want the stud wall layout to dictate where the receptacle will be placed along the countertop, you can make your own design. To do this, put horizontal nailers in between the studs along the countertop. The height of the common kitchen counter is approximately 40 in. at the backsplash. Installing the nailer bottom at 42 in. will place the receptacle just above the backsplash. In this situation, I prefer to install the receptacles horizontally, which makes a more pleasing line with the horizontal counter. You can now plan the distance between the receptacles to match your needs.

For lighting, place a wall-mounted switch at each entrance to the dining/kitchen area. If three-way switching is needed (for example, if the kitchen has more than one entrance), these switches can be fed from a 14-gauge, 15-amp light-duty lighting circuit (the circuit that feeds the smoke detectors). In addition to overhead lighting, consider adding a light at the sink and installing under-counter lighting.

Bathroom

You may think that the bathroom, being so small, would be quite easy to design and wire. No such luck. Next to the kitchen, the bath is the most complicated room to get right. As I did in the kitchen section, I'll first go over what code requires, then offer my Above Code version.

THE NEC BATHROOM Code requires that each bathroom have its own 20-amp GFCI-protected circuit—either a GFCI breaker or receptacle. In the past, this receptacle had to be isolated or, at most, tied in to other bath receptacles (an infinite number of receptacles in an infinite number of baths). As if the infinite number of receptacles wasn't bad enough, the 1999 code made it even worse. As long as the receptacle doesn't tie in to any other bath, code now allows you to load it up with anything and everything in that bathroom (whirlpool tub, lights, heaters, and so on).

Code also requires switched lighting at the point of entrance and a receptacle within 36 in. of the sink lip (the receptacle must not be in an adjoining room). Double sinks do not need two receptacles as long as the sink lip of both is within 36 in. of a receptacle. If the bath doesn't have an openable window, it needs a vent fan. If the fan is in the shower/bath area, it must be listed for wet locations and must be GFCI-protected.

Lastly, no hanging fixture or lighting track can be located over or within 3 ft. of the tub, unless it is more than 8 ft. above the tub.

THE ABOVE CODE BATHROOM

I much prefer a common-sense version of the NEC, keeping the receptacle unto itself. Allowing more loads on a circuit that, as the years go by, takes on more and more high-wattage loads, such as hair curlers and hair dryers, is not a wise decision. I recommend isolating the bath receptacle from all other bath loads. In addition, I also recommend that when there's more than one receptacle, each one should be wired on its own circuit. This may seem like expensive overkill at first, but it isn't. It is quite common to have two people (both spouses or two teenagers) working on their hair at the same time. Two heavy-amperage hair dryers cannot work simultaneously without their own circuits.

Heating a Bathroom

No one likes a chilly bathroom, especially when dripping wet. You can always crank up the main thermostat—if you like a tropical climate and don't mind paying sky-high fuel bills—but a space heater is a better solution.

There are a lot of space-heating options, ranging from portable heaters (most of which are illegal to use in bathrooms) to in-floor radiant systems (wonderful but pricey). I advise my customers not to use overhead fan/heater/light combination units. This type of heater throws heat overhead; in that case, the heat stays near the ceiling or is sucked out by the vent fan. Electric baseboard units are not a good choice, either. They have a tendency to rust, especially when mounted close to a shower or toilet.

I generally recommend a fan-assisted, in-wall heater with an integral thermostat (240-volt with its own circuit). This type of heater circulates hot air, which helps absorb moisture and prevents the temperature from settling in different layers (which can lead to frozen pipes).

To wire an Above Code bathroom, start by bringing in two power cables: one for the receptacle and one for the lights, fan, and so on. If you want a heavy-duty heater, bring in a third. Bring the power cable for the lights into the light switch outlet box by the door. Typically this is a double box—one switch for the fan and another for the overhead lights. The lights by the mirror, if any, are powered by a switch on the wall adjacent to the sink (run a jumper over to it from the door switch box).

The power cable for the bath receptacle is a home run from the service panel directly to the receptacle outlet box adjacent to the sink. Bring in an additional home run for each receptacle. A whirlpool tub, if present, requires its own GFCI-protected circuit starting from a GFCI breaker and terminating at a hard-wired splice at the tub's motor. Don't forget that the tub must be installed so that the motor can be removed, if needed.

Hallways and stairs

Code requires one receptacle if the hallway is longer than 10 ft. Stairs with six steps or more require switched lighting at each end and at a landing that opens to a door. Code doesn't require receptacles on stairs, but I think it's a good idea to install an extra receptacle either at the top, middle, or bottom of the staircase and at one of the landings. This will give you a place to plug in a vacuum cleaner, nightlight, or Christmas lights for the railing.

Bedrooms

Bedrooms require switched lighting via either an overhead light or a switched receptacle. Always go with overhead lighting; otherwise, you will be stuck with a table lamp on that end table at that one spot forever. Once you mark the light outlet box, determine the location of the bed, dressers, end tables, and any other specific piece of furniture. If you think you're an average person and don't need anything other than the few outlets required by code, the wiring is fairly straightforward. But I've never met an average person. I advise my customers to consider everything that they may want in the room and to wire it accordingly. For example, will the room house a computer or stereo system? Also consider track lighting above the bed, overhead lighting, and paddle fans.

Closet lights

You don't have to install closet lights, but if you do, the rules are quite simple. First, always use an enclosed bulb. Second, install the light immediately above the top of the closet door trim. This is about as far away from anything flammable as you can get. Keep all light fixtures a minimum of 12 in. away from all flammables even though code allows 6 in. in some situations. Never install a hanging fixture in a closet.

Dining room

The dining room needs switched lighting at each entrance. The standard rule for placing receptacles applies here. All receptacles within the dining room, kitchen, and pantry cannot be shared by circuits in any other rooms. For example, if a bedroom shares a wall with the dining room, it cannot tap off one of the dining room's receptacles.

Living room

The living room needs switched lighting, preferably at each entrance. The lighting can be overhead or a switched receptacle, but overhead is always preferred. In addition to the standard receptacle placement, it's a good idea to place receptacles on both sides of each window and corner, as well as on both sides of the

Rex's Dream Bedroom

The design of a home can bring a relationship together or tear it apart. Although I don't have the space to go into detail about complete house design, I'll give you some of the more popular electrical designs for bedrooms that can help reduce tension and increase comfort. Remember that no one has to live in a cookie-cutter room—ask for what you want.

Outlets

The rule here is that more is better. It has been my experience that you need a receptacle outlet within 2 ft. of each room corner (one on each corner wall, except where there is a door), and if the room is large enough, another one in the middle. If you have windows and there is no receptacle close to either side, consider installing an extra outlet adjacent to one or both sides of the window. Having an outlet in each corner gives you the flexibility of putting furniture almost anywhere you want. The windowsills can now accommodate electric candles, Christmas stars, and other animated toys you may want to display during the holidays.

Install a receptacle quad on each side of the bed adjacent to the end tables for all your creature comforts: clock, radio, TV, transformer for the laptop computer, complete stereo system, portable CD player, ion generator, sleep machine, window air conditioner, half fridge, fan, VCR, DVD player, and so on. You'll never run out of things to plug in—only receptacles to plug into.

Lights

If you're fairly certain that the bed won't be moved, I recommend installing a switch on each side of the bed to control the overhead or wall (sconces) lighting—that's a four-way switching arrangement. This way, you can dim the lights to whatever mood seems appropriate without leaving the bed. If you have lights below an overhead fan, they can be controlled by remote.

Reading in bed is sometimes a point of contention between spouses. To be able to read at night and not send your significant other out of the room (or worse, have to endure constant suggestions to turn out the light), consider installing track lighting on the wall above the bed and wired or recessed lighting with dimmers in the ceiling. With both installed, the light can be concentrated on just one side of the bed.

Above Code Placement of Bedroom Receptacles and Switches

Window

Receptacle outlets at each corner

Receptacle outlets adjacent to each window

Chest of drawers

Receptacle outlets on both sides of chest of drawers

L

Window

Proposed bed location

Proposed chest-of-drawers location

Wall sconce

Door

Switched overhead lighting

Switched wall sconces

Typical Minimum-Code Receptacle Layout

Window

12 ft.

12 ft.

12 ft.

Window

6 ft.

6 ft.

12 ft. or less

Door

fireplace. Be sure to place extra receptacles for the stereo, VCR, television, and any other electronic gear that may be used in the room. It's a good idea to put the electronic gear on its own 20-amp circuit.

Garages and unfinished basements

Attached garages and unfinished basements are required to have at least one GFCI-protected receptacle and light (an unattached garage does not require these,

unless power is brought in). Garages and basements may have workbenches, so be sure to install a few receptacles at workbench height (around 48 in.). Any accessible general-purpose receptacle must be GFCI-protected, unless the receptacle is for a dedicated purpose, such as a freezer, or is inaccessible, such as a garage-door opener. A dedicated receptacle for an appliance must be a single receptacle unless both parts of the duplex are used.

If the garage shares a wall with a room in the house, never install back-to-back receptacles within the same wall cavity. In a garage, the drywall is a fire barrier. If you cut a hole in the fire barrier in the garage and then cut another hole inside the house in the same wall cavity, you open up a fire entry into the house. It's especially important to avoid installing a panel in this wall, too. This is a common and expensive mistake.

ABOVE CODE The question always comes up as to what is considered finished and what isn't. Officially, "finished" means habitable. So if someone is currently living in or can live in the basement, then it is habitable and the GFCI requirement does not apply. In my opinion, electricity doesn't care who's living where; it's only looking for an easy path to ground, which means that the floor material is the most important consideration. If the basement floor is bare concrete, then the receptacles need to have GFCI protection.

Laundry room

In addition to an electric dryer (which will need its own 240-volt circuit), laundry rooms require at least one 20-amp circuit that cannot be a shared with any other room. Although only one receptacle is required, you'll probably need a few more. Scatter at least two or three around the room. For example, determine where you plan to iron in the room and make sure you have a receptacle nearby. A gas dryer will need 120 volts and can plug into any of the duplex receptacles. You'll also need switched lighting as you enter the room. Put this circuit on any other lighting circuit in the house, but not on the utility receptacle circuit. If the washer and dryer are located in an unfinished basement, the utility receptacle cannot

double as the one receptacle required for the basement.

Attics, crawl spaces, and outside

If there is maintainable equipment in the attic or crawl space, it must be accessible and there must be a switched light at the point of entrance that illuminates the equipment and the path to it. The light fixture should not be a bare bulb. It is too easy to hit your head on the bare bulb and injure yourself and too easy for a hot filament to fall on flammable storage items.

Even if no maintainable equipment is inside, if the attic is accessible via stairs or a foldable ladder, it should have switched lighting. If the switch is located in the attic, place it within arm's reach of the scuttle hole. If it's outside the attic, place the switch reasonably close to the scuttle hole entrance.

You are required to have two outside receptacles: one in the front of the house and one in the back. If you have a deck, consider installing an outlet on both sides of it. These come in handy for extension cords, Christmas lights, cooking equipment, radios, TVs, and so on. If you have a long wraparound deck, consider installing a receptacle outlet every 12 ft. These receptacles must have GFCI protection.

Outside outlet boxes can be flush-mounted on the finished siding or recessed within the wall. Flush-mount boxes must be listed for outdoor use; recessed boxes can be the common nail-on types. In most cases, flush-mounted boxes are easier to install, because you don't have to cut the siding around the box opening. All you need to do is drill a hole through the siding for the wire and attach the box with a few screws. (For more information on installing outlet boxes, see chapter 9.)

Maintaining Structural Integrity

No matter where the cable is run, the structural integrity of the wood must be a prime consideration. That is, do not drill a large number of large-diameter holes throughout the framing member to pass cable through. Local codes dictate whether or not the structural integrity of any framing member—stud, plate, or joist—is being violated. If you have any doubts, it's a good idea to check with the local inspector or lead carpenter before drilling or cutting. Some common codes for retaining structural integrity are illustrated below.

Notching and Boring Studs

Drill the hole exactly in the center, with the hole edge a minimum of ¼ in. from the edge of the stud; otherwise, ¹⁄₁₆ in. from the steel plate is required.

The hole should not exceed 40% of the stud width, or 1⅜ in.

The hole should not exceed 40% of the stud width, or 1⅜ in.

The notch should not exceed 25% of the stud width, or ⅞ in.

The notch should not exceed 40% of the stud width, or 1⅜ in.

2×4 bearing wall   2×4 nonbearing wall

Drilling Multiple Holes

OK

OK

Violation

Protecting Cables

When a stud is notched or bored, code requires a ¹⁄₁₆-in. steel plate to protect the cables from screws and nails.

A steel plate is required when the distance is less than 1¼ in.

Steel plate

Cable

Notching and Boring Joists

Holes must be minimum of 2 in. from the joist edge and cannot exceed one-third of the joist width.

A joist can only be notched on the outer third of its length, not in the middle third. Notches cannot exceed 25% of the joist width.

2 in.

If you plan to leave something plugged into the receptacle for an extended period of time, such as holiday lights, the lid must be a watertight-while-in-use lid—one that maintains its watertight integrity while the cord is plugged into it.

Required outside lighting is limited to grade-level access points for all doors—excluding garage doors. The lighting can be indirect, such as an overhead floodlight, or a light right at the door. Neither light has to be switched.

Pulling Cable in New Construction

After you have all the boxes mounted, it's time to pull the cable. You can route cable through the crawl space or the attic and through, or over, studs and joists. For neatness and safety, try to keep the home runs along the outer walls.

Don't bundle groups of cable together for long runs. It may look neater, but if bundled tightly together for 2 ft. or more, the heat generated within the bundle may damage the wire's insulation. This means that you may have to lower the amount of current flowing through the wires (called derating) to lower the temperature and thus protect them. For instance, for a bundle of 10 to 20 conductors (five to ten 12-2 w/g cables), a 12-gauge cable must be fused at 15 amps, a 10-gauge cable at 20 amps (the NEC gives guidelines for derating conductors). To avoid having to make these adjustments, don't bundle the cables together. It will make your life easier.

I once went to a job that had failed inspection, and the owner was paying me to fix another electrician's problems. It was the neatest job I had ever seen—even the stranded neutral had been braided like a girl's pigtail. The electrician ran all the wires—10 cables (20 or more current-carrying conductors)—through the same large holes and bundled everything tightly with tie wraps for the entire run. It looked great, but it failed inspection because the cables, especially the ones going to the kitchen and laundry, could not be derated.

Through basements and crawl spaces

If a basement will not be finished, 8-gauge and larger NM cable can be run across the bottom edges of the floor joists, or you can nail a running board under the joists and staple the cables to the board. Cables smaller than 8 gauge that are run across the bottom of the joists must be stapled to a running board.

If the basement will be finished (either now or in the future), run the cable through the floor joists or along one of them. When running cable along the joist, staple it every 3 ft. to the center of the face to avoid the nails and screws that will be driven from below when the ceiling is installed. When boring holes to run cable through joists, follow the guidelines shown in the sidebar on the facing page. Crawl spaces have no rules for running the cable along the joist edges because you can't trip over them.

Here's how to pull the cable from the main panel, through a basement or crawl space, and to the outlet.

1. Lay the box of cable in front of the main panel and unroll the approximate amount you'll need. Try to unreel the cable from the outside of the roll so that it will remain flat, as opposed to taking it from the center and having it loop in spirals. Next, feed one end of the cable through the lubricated slot in the bottom plate.

Unrolling Cable Correctly

Before you pull the cable into the walls, unroll the cable as you would a bait casting reel. Wrap one end around a stud and walk backward, spinning the cable roll in your hands and unrolling enough cable so there is slack to pull through the studs. Doing it this way allows a much easier pull because the cable is flat. Pulling the cable out of the center of the box, as it is designed to do, makes the cable spiraled and harder to pull through the drilled holes.

2. Pull the cable through the holes in the joists or across the joists (the method you choose depends on whether the basement will have a finished ceiling) to the wall cavity in which the cable will terminate.

3. Drill through the bottom plate of the interior wall. Pull enough slack to the location of the receptacle or switch.

4. Staple or attach the cable wherever needed in the basement (along the face or edge of the joists or on the running board). Attach it every 3 ft., not the code minimum of 4½ ft.

5. At the main panel, cut the cable with enough slack to go from the large hole in the bottom plate to the top of the main panel.

6. Mark the cable to indicate where it goes.

Through attics

Once you've planned the wiring route through the attic, you can begin to pull the cables. Here's how to do it (this method can also be used for joists or trusses).

1. Lay the box of cable in front of the main panel and unroll the approximate amount you will need from the outside of the roll. Work out the coils so that the cable lies flat. Next, feed one end of the cable up and over the first ceiling joist. Don't insert the cable through the large hole in the double top plate above the panel yet. Doing so will make it much harder to pull it over the top of the joists and could damage the cable because of the sharp 90-degree turn it has to make immediately above the top plate.

2. Pull the cable over the joists to the stud-wall cavity you need to access. Wiring home runs (the cables from

Running Cables in an Attic

Wires shorting to each other

Footprint of the last person in attic

2×4s on edge nailed onto the side of a 2×6 running board

Joists

Cables

Light fixture

Light fixture

Staple

the main panel) along an exterior wall allows you to keep the wiring out of harm's way. To get cable to an interior partition wall, simply pull through enough slack to make the run along the edge and then along the face of the ceiling joist to the cavity.

3. Once you reach the correct wall cavity, drill through the top plate and pull enough slack to get to the intended receptacle or switch.

4. Insert the cable through the hole in the top plate and pull the slack all the way through to its intended location. In the attic, staple the cable wherever needed (along the face or edge of the joists). Attach it every 3 ft., not the code minimum of 4½ ft.

5. At the main panel, cut the cable with enough slack to go from the large hole in the double top plate above to the bottom of the main panel.

6. Insert the cable from the attic through the double plate and into the main panel.

7. Mark the cable to indicate where it goes.

Wiring through Trusses

If trusses are used in construction, be wary. Do not drill or cut trusses to run cable (unless approved by the truss manufacturer and building inspector), because it could ruin the engineered specs and void the manufacturer warranty.

A floor truss has an open area around 12 in. to 18 in. tall. The space is perfect for running cable across without drilling through the trusses or installing a running board (it's okay to staple wire to the edge of the truss members). Attic trusses, however, have larger open spaces and may be used in accessible areas. If you need to run wire across the trusses, install a running board on top near the truss edge and staple the cables on top of the board. (The truss edge is located above the top plate along the outside wall.)

Wiring over and through Studs

From attic

1¼ in. minimum on each side of cable

Drill a hole 6 in. to 8 in. above or below box.

A staple is required within 12 in. of metallic box and within 8 in. of nonmetallic box.

Along and through studs

When wiring vertically along a stud, keep the cable to the center of the stud (maintain 1¼ in. of space on each side of the wire) and staple it every 2 ft. to 3 ft. If there will be several cables run in a very narrow location, most inspectors will allow two cables under one staple. But if you need to stack more cables, or if there isn't enough room under the staple, stack the cables through special plastic brackets found at electrical supply houses. If you need to drill through the studs, follow the guidelines on p. 126 to maintain the structural integrity of the wood. In addition, be sure to drill at least 8 in. to 10 in. above or below the receptacle or switch location, because code requires the cable to be stapled to the stud within 12 in. of

Using a height gauge helps you drill all the holes at the same height for the cable runs. This gauge is a bamboo stick with a notch in one end in which the spade bit shaft rests. Remove the stick after the bit punches into the wood.

a metal box and within 8 in. of a nonmetallic box.

Make sure that you drill the studs at the same height on horizontal runs. Having the cables zigzag up and down in the stud wall is ugly, wastes cable, and damages the jacket during the pull. There are many ways to keep the run straight. Experience is the first; most longtime pros can drill holes consistently in a straight line. Most beginners can't. The easiest way I've discovered is to use a little cane stick with a notch cut in one end. Just lay the spade bit in the notch and drill all the studs close to the same spot each time.

Around windows

Running cables around windows takes some forethought. Your first idea may be simply to drill every stud that gets in your way, including the triple stud on both sides of the window. But this method requires drilling six to eight

Running Wire through Studs

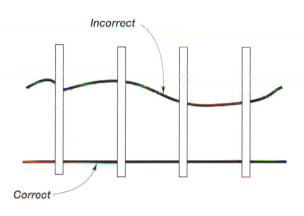

Incorrect

Correct

times. Normally, it's better to take the cable up through the double top plate, through the attic along the exterior wall, and then down again through the top plate to the wall cavity on the opposite side of the window. If you prefer, run the wiring through the basement. With the latter two methods, you only need to drill twice (see the drawing below).

Wiring around a Window

Through attic

Through studs

Through basement

To get cable around a window, run it up through the attic, through the studs below the window, or through the basement or crawl space.

Wiring around a Door

To get wire around a door, run it through the attic, through the basement or crawl space, through the cripple studs above the header, or through the shim area of the door.

Through cripple studs

Through attic or ceiling joists

Through shim space

Through basement or crawl space

Once in a while you may encounter a window that runs from top to bottom with no place to run the cable overhead due to an exposed-beam design. If there is no crawl space, your first option is to run the cable through the shim area between the header and the window frame. If that is nonexistent, you'll need to make way for the cable using a router or a circular saw. Set the bit or blade to a depth of no more than ½ in. (⅜ in. is better) and create a channel across the header to lay the cable. Once the cable is in, cover it all the way across with a ¹⁄₁₆-in. steel plate to protect it from nails and screws.

Around doors

Getting cable around doors also takes some forethought, but the logic is the same as getting around windows. The easiest route is to go up the stud and through the attic or ceiling joists, or you can go through the basement or crawl space. In these situations, you only have to drill twice. Another option is to run

the cable through the cripple studs above the header, but this method requires a lot of drilling, and sometimes the door header is so high that there's not enough room over the door. Yet another option is to run the cable through the shim area between the door frame and the header and protect the cable with a ¹⁄₁₆-in. steel plate.

Once in a great while you'll encounter a house with an exposed-beam design and doors with solid headers with no shim room. Whenever there is absolutely no place to run a cable, use a router or a circular saw to create a channel in the header. And don't forget the ¹⁄₁₆-in. steel plate.

Boxed corners

Boxed corners are always problematic. If the boxed corner looks like a major problem, try to run the cable over and around it. If you cannot go around, which is rare, note whether the box is hollow or solid.

HOLLOW BOXED CORNERS   A hollow boxed corner is easy to get through. Using a 1-in. spade bit, drill from both sides at the same level. Then it's simply a matter of fishing the cable through the holes. If a stud is in the way and you can't get the drill into the wall cavity (and if the carpenters don't mind), remove it temporarily. Carefully cut the nails at the top and bottom of the stud using a reciprocating saw with a fine-toothed blade, then lift the stud out. Cut parallel to the top and bottom plates, and don't ruin the stud by bashing it out with a hammer—the carpenters may massacre you. When you're finished, nail the stud back in place.

SOLID BOXED CORNERS   To get through a solid boxed corner, first remove

How to Cut through 15½ in. of Wood with a Short-Shank Spade Bit

I prefer to use a short-shank spade bit to make a deep cut, because a long spade bit tends to snap in tight holes, it's hard to get the cuttings out of the hole with large bits, and an auger bit tends to jam in the wood during deep drilling. Although this example uses a spade bit, you can use the same trick with auger bits; just start with a large bit and taper to a small one.

A common ¾-in. spade bit can cut through 4½ in. of solid wood. A ¾-in. spade bit and a ⅝-in. spade bit with a 6-in. extension can cut through 9½ in. of solid wood. Drill to the full depth with the ¾-in. bit and then finish the hole with the ⅝-in. bit and extension.

Use a ½-in. spade bit with a 12-in. extension to drill through 15½ in. of solid wood.

15½ in.

6 in.   5 in.   4½ in.

½-in. bit   ⅝-in. bit   ¾-in. bit

A ¾-in. spade bit can go into wood approximately 4½ in.

A ⅝-in. spade with a 6-in. extension allows a total drilled depth of 9½ in.

A ½-in. spade with a 12-in. extension allows a total drilled depth of 15½ in.

Fishing Cable through a Solid Corner

1. Insert fish wire into the hole.

2. Hook fish wire with another single wire and pull it through.

3. Lubricate the hole and pull the hook, fish wire, and cable through the hole.

Chiseling and Drilling through a Diagonal Corner

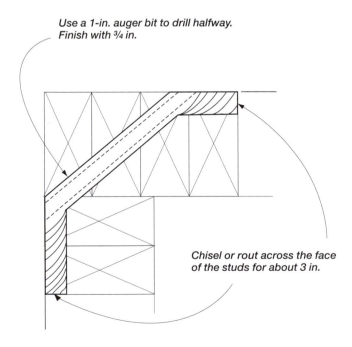

Use a 1-in. auger bit to drill halfway. Finish with ¾ in.

Chisel or rout across the face of the studs for about 3 in.

any studs that are blocking your access with the drill. Next, drill a 1-in. hole a little over halfway into the corner. Be careful while drilling. The drill is working hard and could buck or stall, and there are a lot of nails in a solid corner. You can use a ⅜-in. drill or even a battery-operated one, but it will take longer and put a lot of wear and tear on the tool. Do not use a pistol-grip ½-in. drill because if it stalls in the wood or hits a nail, the drill could break your wrist. If you use a spade bit for this operation, make sure it is very sharp. If you use an auger bit, use one that can go through nails, such as a Greenlee Nail Eater II. Once the hole on one side is complete, drill from the other side until the holes meet.

If the entire corner is so thick with studs that it isn't feasible to drill, then you have two choices. You can chisel or use a router to open a ½-in.-deep pathway along the stud face. Or you can open the pathway partway, drill through the corner at a 45-degree angle, and then continue the pathway. I prefer the latter method, because it doesn't put the cable in a 90-degree bend around the outside corner. If it's an inside corner, you have no choice but to have the cable in the sharp bend. Remember to attach the steel guards to protect the cable.

Once the drilling is done, fish the cable through, as shown in the top drawing on the facing page. This is easier said than done. First, be sure to clean all the wood shavings out of the hole before you start, and put some cable lubricant into the hole to make the cables slide easier. Insert a single wire through one hole to the very end. The single wire serves as a fish wire. Then, using a stiff wire with its end formed into a hook, reach from the hole on the other side and hook the fish wire. At the same time, pull and feed the fish wire until it comes out the other side of the corner. If it falls off the hook, reach in with some needle-nose pliers, grab it, and pull it the rest of the way out.

Drill a hole (at the same height as the holes in the boxed corner) through each stud you had to remove earlier and nail it back in place. Pull the cable through the boxed corner and all appropriate studs. Be careful not to damage the cable as you pull it through the 90-degree bend. It's wise to insert a second fish wire through the hole as the cable is being pulled through, just in case you decide later that a second cable needs to go around that corner, too.

Concrete walls and floors

Getting cables through concrete without the proper tools is very difficult. I use a rotary hammer with expensive carbide-tipped drill bits to drive through solid concrete. However, a simple star drill, or even just a hammer, will punch a hole through hollow concrete block.

There are a few methods for running cables through basement walls of cinder block or solid concrete. The least desir-

able way is to run them through the inside of the walls, with the boxes installed in cavities broken into the blocks. This, of course, cannot even be considered with poured walls. The most common method is simply to nail and glue pressure-treated boards where the conduit and boxes will be installed. Then attach the conduit, cables, and boxes to the wood. Otherwise, you'll wind up drilling and inserting some type of anchor for every box and anchor point for the conduit—and that's a lot of drilling and anchors to install. If the basement will have framed walls flush with the concrete walls, you can run the wiring through the framed walls.

Never pour concrete around a wire or cable; always install conduit for the wiring. If outlets need to be in the middle of a concrete floor, which may be the case in a garage or shop, run the conductors through conduit before the pour and leave the conduit sticking out above the floor. Avoid installing boxes flush with the floor because water could pour into the outlet or trash could collect there.

Wiring outside

You do not need to use UF cable every time the cable leaves the house. If you can keep the cable out of direct rain and sun, such as under the roof of a porch, you still can use NMB cable. However, if you run a cable under a deck with no overhead protection and the cable will receive rain or runoff from the deck, you need to use UF cable.

As a rule, bury electric cable at least 24 in. deep. This not only keeps you within code, it also prevents you from cutting the cable when you dig in your flower bed. If you know you are in an area that won't be dug up and are burying a GFCI-protected 120-volt line (for example, from the house to a pole light), you can raise the depth to 12 in. Tables

WORK SAFE
WORK SMART
THINKING AHEAD

If you can, predrill the studs that will make up a blocked corner before they're installed by the carpenter. Then all you have to do is just pull the cable.

WORK SAFE
WORK SMART
THINKING AHEAD

When using an auger bit to drill deep holes, wax the sides of the bit. The wax helps keep the bit from jamming in the hole.

within the NEC give depths for varying situations.

Remember that the conductors, even UF, must be protected from physical damage. That means that you can run the conductor under the deck without physical protection, but as soon as it rises above the deck, it must be in conduit to a high enough point that it is no longer in danger of being damaged (normally around 8 ft.).

Routing Wires in Renovation Work

Renovation is not easy. Rewiring a house is very labor-intensive, sometimes taking half a day just to replace or add a single receptacle or switch. I try to run wires so that I make as few holes as possible in the walls. The easiest and cleanest routes are through the attic and basement, but these routes are not always available.

One of the first things you need to learn is how to find studs behind a finished wall. I normally use a stud sensor. This may or may not work on thick plaster walls. If all else fails, you can try drilling $\frac{1}{32}$-in. or $\frac{1}{16}$-in. holes or drive in a finish nail until you feel solid wood. Another option is to drill a $\frac{1}{4}$-in. to $\frac{3}{8}$-in. hole, insert a wire bent to a 90-degree angle, and then rotate the wire until you bump the stud. If the trim can be removed, cut into the wall behind the trim to locate the studs. Another trick is to remove a switch or an outlet cover and see which side of the box is nailed to the stud, then go every 16 in.

Opening a finished wall

There are several ways to open a finished wall. You may be tempted to use a jigsaw or reciprocating saw to open a wall, but I don't recommend it because of the high probability of cutting the existing wiring or even the plumbing in the wall. If you know for a fact that there are no wires or plumbing within a particular wall, then use a jigsaw to open the wall. But make sure the tool is equipped with a blade with many small teeth to reduce pulling, cracking, and splintering the finished wall. To keep the saw's bottom plate from scuffing the wall, cover the plate with tape. Try to cut through just the wallboard. Adjust the depth of cut so that it just cuts through the wallboard. In many cases, I prefer using a sharp utility knife or a keyhole saw. A slow, manual cut is much safer and neater than a fast cut with a power tool.

Plaster on lath presents special problems. No matter how careful you are, sawing or drilling holes frequently splits the plaster or produces horizontal cracks in the wall. Old molding has the same problems. If the molding is unique, you may not be able to get a replacement piece if it gets damaged.

When drilling into walls, any drill will do, but it's better to use a cordless drill than an AC-powered one. If the drill bit of a cordless drill cuts into a hot wire, the drill, being mostly plastic, won't electrocute you. Never use an old, ungrounded metal-handled drill, unless you have a death wish. If you use a corded drill, make sure it is powered from a GFCI circuit or a GFCI-protected extension cord.

Wiring through the basement or crawl space

To run new cables through the crawl space or basement, first locate where they will run to the new receptacle or switch. The simplest way to do that is by first drilling a $\frac{3}{16}$-in. pilot hole in the floor in front of the receptacle or switch in the wall. (If there is carpet on the floor, use a utility knife to cut a small X where you'll

WORK SAFE
WORK SMART
THINKING AHEAD

Be very careful when pulling several wires through conduit, because the individual wires can get tangled easily. As you assemble the conduit, run a pull wire through each section. Once the conduit is assembled, tape all wires to the pull wire, smear cable lubricant all over the wires, and pull them through.

be drilling. This keeps carpet strands from getting caught in the drill bit.) Once you've drilled through, either leave the bit in the hole or pull the bit out and insert a piece of wire through the hole to the crawl space, which will give you a reference point in the crawl space. Move about 2 in. to 3 in. from the reference point, find the bottom plate, and drill through the subfloor and plate into the wall cavity. To help you find the bottom plate, look for nails extending through the floor that attach the plate to the sub-floor. If you're replacing or rewiring an existing switch or receptacle, you should be able to see where the old cables go through the subfloor and into the wall cavity. Be careful not to hit your head on the nails, and wear safety glasses to protect your eyes from falling debris while you drill.

Wiring through a finished wall

Whenever a switch or receptacle is being relocated or added, you may have to route the new wiring through a finished wall. The first step is to de-energize all circuits within the wall. Bring in additional light, because de-energizing the circuit may also remove power to the lights in the room. Then locate the new box, trace around it on the wall, and cut out the hole for it. The best choice for remodeling work is a cut-in box because it has flaps that easily attach to the drywall. (If you choose another type of box, it may have to be nailed or screwed to the stud, which means removing a lot of the drywall.) If you're starting from an existing box that needs to be removed, another option is to use a Makita mini-reciprocating saw to cut the nails of the existing in-wall box and remove it. If the drywall is being opened for a new box, the box should be located about 1 in. from the stud toward the center of the wall cavity. Do not cut

Creating a Reference Point in a Crawl Space

Location of receptacle

½-in. drywall

Trim

Insert the locator wire through the pilot hole for a reference point in a crawl space.

Bottom plate

Floor

Subfloor

Move 2 in. to 3 in. from the reference point and drill through the plate.

the hole for the box so close to the center of the wall cavity that you won't be able to reach the adjacent stud with your drill.

To run cable horizontally through several studs to a new box, use a utility knife or a keyhole saw to open a 2-in. channel through the wallboard from box to box. Drill through the studs and pull the cable through to the new box. If you're relocating or adding a box in the wall cavity adjacent to an old box, simply remove the old box from the wall and drill at an angle through the stud to the adjoining wall cavity. Make sure the hole is above or below the location of the old box. If the old box is staying, it won't fit back into the hole because the new wiring will be in the way.

When you install a new receptacle in a room and there is one in an adjoining

Wiring through a Finished Wall

Right Way

Drill the stud through the box hole.

Open a 2-in. channel in the wallboard from stud to stud, then drill through the studs.

The hole for a cut-in box is about 1 in. from the stud and below the hole in the stud for the cable. The center of the channel is level with the top or bottom of the hole.

Fish cable across here.

Studs

Wrong Way

With this location, the box will not fit because the cable will be in the way. The box hole should be mounted higher or lower than the drill hole or moved farther from the stud.

The box hole was cut too close to the stud (if a cut-in box is used).

There was no need to open a channel this long. This particular wall cavity didn't need to be opened at all to get the cable through.

Studs

Routing Cable behind Baseboard Trim

To run cable through the studs behind existing wallboard, remove the baseboard section. Cut out the wallboard from the center of the stud nearest the existing box to the center of the stud nearest the new box.

Existing receptacle box

New receptacle box

Cut out the wallboard behind the trim.

Remove the trim.

room sharing the same wall cavity, power the new receptacle through the existing one. Turn off power to all circuits in that wall. Mark the location for the new box and open up the wall. Be aware that there are wires inside the wall that you don't want to hit. Do not place the boxes back to back, because the wall probably won't be wide enough. Instead, move the proposed box slightly to one side of the existing one.

Remove the existing receptacle from its box and find which side of the box is nailed to the wall stud. You have 16 in. center to center from that stud to the next one in which to locate the new box. If the old box is plastic, open up the cable slot with a screwdriver so that the new cable will go through smoothly. If the old box is metal, loosen the cable clamps and open up one of the knockouts or cable openings with a screwdriver. If you remove a knockout, install a connector to protect the new wire from chafing on the metal. Push a new cable into the old box, giving yourself 6 in. of cable out the box front. Cut the cable to its approximate length and feed it through the hole for the new box, leaving 6 in. Then install the new box.

Behind baseboard molding

You can route cable through the wall behind existing baseboard molding, but only when the molding can be easily replaced if it gets damaged. Do not try this method on old baseboard that is irreplaceable; it may split or otherwise become damaged as you remove it. This method is most commonly used in situations when an original electrician left out a feed wire to a switch or when an owner wants to add a receptacle or fixture after the finished walls are up.

The first step is to de-energize any circuits within the wall. Make a pencil mark along the top of the wall right

Never use a screwdriver to remove a staple or you'll damage the cable. If you can't get the staple out with electrician's pliers or the end nippers, try an old-fashioned nail puller. Simply put the staple in its teeth and pull back on the handle.

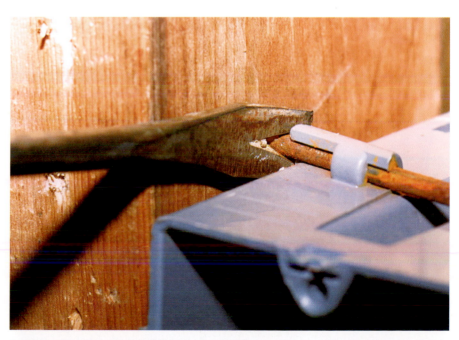

A trim nail puller may look like just a bent screwdriver with a forked tip, but it is the perfect tool for removing an integral nail box from a stud without damaging the box.

My trim pull kit consists of a 6-in. prybar (with the edges ground down), a piece of cardboard, and an outlet box with one side cut out. The box serves as a pivot point and keeps the wall from getting dented. The cardboard keeps the box from marring the wall. Once this prybar pulls out the trim ¼ in., I replace it with the larger prybar.

above the trim. Pop the molding off as carefully as possible. Use a utility knife to cut any paint holding the molding to the wall and use a nail set to drive the trim nails through the wood, which will free the trim board. If you prefer, you can use a wide chisel or prybar to pull the trim off.

With the trim removed, open up a channel in the wallboard using a keyhole saw or a utility knife. Do not remove any finished wall above the mark. Using a long bit, drill ½-in. or ⅞-in. holes in the studs and pull the horizontal section of cable first. Wear safety glasses and be careful not to hit any nails. Run a fish wire from the proposed box hole, tape the new cable to it, and pull both into and out of the hole. Remove the fish wire, run the new cable into the box, and insert the box into the wall. If you're bringing cable into an existing receptacle box, run a fish wire from the bottom of the box to the opened area behind the trim. Tape the new cable to the fish wire and pull both up into the old box.

Around doors

Routing wiring around a door is normally done when a wall switch was installed on the wrong side of the door frame or when a new switch is being added and power is coming from the receptacle on the opposite side of the door. The procedure also works for windows. As always, first de-energize the circuit you're working on and any that may be within the wall. Remove the trim. Then, using a sharp chisel, notch the shims and anything else that may prevent the cable from going around the door. Route the cable from the receptacle behind the baseboard, if possible, then up and around the door, squeezing it in around the notched shims. Staple the cable to the shims and protect it where necessary with steel plates. If large sections of the cable are exposed, run it through steel conduit that is at least $\frac{1}{16}$ in. thick.

If a switch will be mounted adjacent to the door, open a hole for the switch box about 6 in. to 8 in. from the jamb. To get the cable from the shim area to the switch box, drill through the door frame from the switch-box side. Insert the cable through the frame and into the box, then install the box.

Through the ceiling or attic

Wiring through the ceiling or attic is normally done when the wiring to a ceiling light is being replaced or when a new ceiling light is being installed. As always, de-energize the circuit to the light and any other circuit in that wall. If the old light is present, remove it and the old box. If the light is a new installation, cut out the hole for the fixture box. Be sure to have the box in hand so that you know what size hole to cut (I'll talk more about installing a ceiling light in chapter 11). In this situation, it's also important to run the wiring parallel to the ceiling joists so that you don't have to open up

Routing Cable around a Door

If there is no other route, run the cable around the existing doorway. Remove the trim around the door and cut away some of the drywall to expose the shim area. Chisel a little from each shim, lay the cable between the jamb and the frame, and staple it to the shims. Cover each exposed section with a 1/16-in. steel plate.

Remove trim from around the door.

Stud

Drywall

Cable

Jamb

Staple

Shim

Remove sections of baseboard trim here.

The steel plate protects the cable.

the entire ceiling to run the cable through the joists.

Next, remove the baseboard trim, cut a channel in the wallboard, and drill through the studs. Cut holes in the wall for the switch box and two slots in the wallboard, one at the top of the wall and the other in the ceiling. The holes must be big enough for a drill to get into them. Drill a 1-in. hole through the top plate with a right-angle drill (remember to wear safety glasses). Run the cable from the receptacle to the new switch box to the ceiling fixture using fish tape or other method, and then install the new boxes.

Routing Cable for a New Ceiling Light

New ceiling box

Ceiling joist

Double top plate

New cable

New switch box

To the light fixture

Existing receptacle

Staple here

New cable

CHAPTER 7

Fuses and Circuit Breakers

OFFICIALLY, fuses and circuit breakers are called overcurrent devices because they provide overcurrent protection. That is, if the circuit pulls more current than the device is designed for, the fuse or breaker will open the circuit. Located in panels and subpanels and sometimes at the point of use (some appliances have built-in fuses), overcurrent devices are designed to protect the house wiring from overloads, short circuits, and ground faults.

An overload occurs when a load placed on a circuit is greater than the rating of the overcurrent device, such as a when a washing machine's motor goes bad and starts pulling 30 amps instead

Cartridge fuses (above) come in two common sizes for use around the home. The larger one is used in circuits for mains (normally 60 amp) and stoves (normally 40 amp) in old-style fuse boxes. The smaller one is normally for point-of-use locations, such as water heaters and old electric stoves. Inside a cartridge fuse (left) you'll find a fire-resistive material around a metal bar element. The size of the element determines the amperage of the fuse.

of 20. A direct short means that the hot and neutral wires are either directly or indirectly touching, such as when a hot wire comes loose in an outlet box or when a nail is driven into a wire. A ground fault is a short in which the hot wire touches ground. For example, a ground fault occurs when the hot wire of a drill touches the drill's metal case. If overcurrent protection is not in place during any of the above problems, the excessive current flowing through the wire produces enough heat to damage the wire's insulation or jacket, and it may start a fire.

There are many types and classifications of fuses and circuit breakers. Entire books have been written on this subject alone. In this chapter, I focus on the most common fuses and breakers that are used in the home.

Fuses

Although today's fuses provide an excellent source of overcurrent protection, it wasn't always that way. They, like circuit breakers, had to go through years of development. Early fuses were simply pieces of small-diameter copper wire connected in series in the circuit. Because the wires were small, they had a high resistance, causing them to overheat and eventually melt or explode when exposed to excessive current. Although the system worked, it was both a fire hazard and a hazard to workers. The copper wire fuse evolved into a lead-alloy fuse, but it still had the same exploding problems. The next invention was a zinc element inserted in a fiber tube filled with fire-resistive powder.

The problem with old fuses—those prior to 1940—was that they took too long to open. This delay caused massive damage to the circuit as a result of the huge current buildup during a fault. Today's fuses are much more sophisticated than those of yesteryear. Modern fuses blow quickly, which cuts the current off in an instant—this is much faster than a circuit breaker operates. The two common fuses we use today are cartridge fuses and plug (sometimes called glass) fuses.

Cartridge fuses

Cartridge fuses (designated Class H by UL) are the oldest type of fuses still in use today. These are one-time-use items. In the old days, you could unscrew the ends of a fuse and replace the element. Today, once the element is blown (melted), you have to replace the entire cartridge.

The two most common cartridge fuses are the ferrule and the knife blade. A ferrule-type fuse protects up to 60 amps. It consists of a fiber tube with two metal end caps that make contact with the circuit. Inside is a zinc element surrounded by fire-resistive powder. A knife-blade cartridge fuse is a ferrule-type fuse with flat metal tabs sticking out from the end caps. It protects circuits above 60 amps and was used as mains in old 100-amp fuse boxes.

The common problems encountered with cartridge fuses are not in the fuse itself but in the fuse holder or fuse box.

A knife-blade cartridge fuse allows a better connection for higher amperages (60 amps and greater).

Burned areas normally signify a loose connection on the fuse holder or within the panel.

The clips that hold the cartridge fuses sometimes get pulled apart over years of changing fuses, and the rivets that hold the fuse clips onto the fuse box loosen over time. Both of these conditions result in the fuse—and sometimes the entire fuse box—arcing, burning, and overheating. Eventually, the heat destroys the fuse by disintegrating the fiber shell. If you can't figure out why a fuse keeps blowing, the fault may lie with the panel. If that's the case, you must replace the entire panel; it cannot be fixed.

Plug fuses

A typical plug fuse has what is called an Edison, or standard, screw base. They are available in ½, 1, 2, 3, 5, 6, 8, 10, 15, 20, 25, and 30 amps, all of which are rated at 125 AC volts. Plug fuses are further subdivided into general-purpose and time-delay fuses. General purpose is the

common, fast-acting fuse. Time delay is used for circuits that protect motors (motors have a high start-up current). Standard-base fuses can be used only as replacements in existing installations that show no evidence of tampering or overfusing.

The problem with standard fuses is that they all have the same type of threaded base. It's easy to substitute one amperage fuse for another and thus create a fire hazard. This design flaw led to the development of type-S fuses, which are basically the same as standard fuses but with slightly different threaded bases for the various amperages. Type-S fuses can also be identified by their nonmetallic threads. These fuses are allowed in new construction; standard fuses are not.

ABOVE CODE For older boxes, install S-type Edison-based adaptors. These adaptors, which screw into the fuse box, prevent someone from inserting the wrong-size fuse into the base.

Old-style main fuse boxes

Old-style 60- and 100-amp fuse boxes continue to provide overcurrent protection in many older homes throughout the country. A 100-amp panel may still offer satisfactory service as long as it hasn't been hit by lightning or been over-

These are Edison-type plug fuses. No matter what the amperage is, they all share the same thread pattern.

Type-S fuses have specific thread patterns for various amperage fuses. The 30-amp fuse (right) cannot be screwed into the fuse socket of the 15-amp fuse (left).

loaded and damaged (look for melted areas around the box and for any areas that give off excessive heat). However, 60-amp boxes should be updated to provide higher amperage, since most of those panels were designed with only four plug-in fuses. You can't power a modern house with only four circuits.

At the time, 60-amp boxes were adequate for the loads that were placed on them. But as the years went by, circuits and appliances were added, overloading the circuits and eventually blowing the fuses. Instead of upgrading the service, many people simply replaced the proper amperage plug-in fuse with a higher-amperage one (usually 30 amps). Thus, wiring that needed 15- and 20-amp protection became overfused, allowing excessive current to heat the wiring and kill the fuse box or start a fire. If you have a fuse box in your house, check the plug-in fuses for any 30 amps (they are usually green). If you find any that do not go to a dryer or water heater, replace them

immediately with the proper size fuse. If in doubt, install 15-amp fuses. Even better—upgrade the service. It should go without saying that modifying a fuse to make it work when it is already blown is a disaster waiting to happen.

Circuit Breakers

Circuit breakers (officially called molded case circuit breakers, or MCCBs) are the most common type of overcurrent protection used today. Simply put, a circuit breaker senses and measures overcurrent in a circuit and opens the circuit. A breaker is designed to be ambient compensated. That is, as air around the breaker heats up, the breaker senses this heat and derates itself. Thus, if a 15-amp breaker is sandwiched between two breakers generating a lot of heat, the 15-amp breaker may kick off at 12 amps.

Most breakers use a two-part system to protect the circuit: one system for mild overloads and another for severe short circuits. In a mild overload, a bimetal

Overfused wiring is quite common in old fuse boxes. Instead of 15-amp fuses, someone has installed two 30-amp fuses (green) and one 20-amp fuse (red). By the time the fuse opens the circuit, the 14-gauge wire's insulation will be damaged and could potentially start a fire.

Checking Fuses

Sometimes you can inspect a fuse just by looking at it. A cartridge fuse may be blackened and distorted, the contact terminals may be arced and burned, or the body may have disintegrated. With plug-type fuses, look at the window to see if the element is broken or blackened.

Many times a fuse element looks good when it isn't. As a double check, use a multimeter or a similar device. A multimeter uses two probes to check a fuse. It sends a low-amperage current into one end of the fuse (via one probe); if the current comes out the other end of the fuse and enters the second probe, the fuse is considered good. The multimeter indicates this by a tone, light, or numbered reading, depending on the device.

The multimeter above has two types of testers: a continuity tester (where it is set now) and an ohms, or resistance, tester (two positions to the right). To use the device, place a probe on each fuse contact (it doesn't matter which probe goes on which contact). The continuity tester sounds a tone and the resistance tester provides a resistance reading (close to zero). If you do not receive an immediate continuity indication, scratch the probe into the contact's metal to bypass any corrosion.

To do a continuity check on a cartridge fuse, place a probe on each of the fuse's metal end caps. Sometimes the caps come loose as the fuse body distorts.

To check a knife-blade fuse, place a probe on each blade.

A type-S fuse needs a probe on the center terminal and a probe on the metal band below the threads.

A plug fuse's terminals are the screw threads and the center terminal. Place a probe on each one.

strip, called a thermal trip, heats up and bends backward, eventually tripping the breaker. For severe short circuits and ground faults, an electromagnet bolsters the bimetal strip, pulling it back even faster for an almost instantaneous trip. When the breaker trips, it opens to a position halfway between on and off. After you have identified the problem on the circuit and remedied it, you can turn the breaker to full off and then back on again. Never hold a breaker on when it is trying to trip off.

Most breakers for the home are designed for copper and aluminum conductors and for wire insulation maximum temperatures of 60°C and 75°C. The temperature rating and CU/AL should be printed on the breaker. Most are also labeled as HACR, meaning that it can be used for heating and air conditioning equipment.

In addition to other specs, every circuit breaker has an ampere, voltage, and short-circuit interrupting rating. The ampere rating is on the breaker handle; it indicates the maximum current the breaker can carry without tripping. Standard residential breakers are rated at both 120 and 240 volts, since those are the common voltages within a panel. The short current interrupting rating is based on the maximum amount of fault current available from the utility. Thus, if the utility transformer generates 10,000 amps of current, the breakers in the residence should be rated to at least that amount of current. Residential breakers have ratings ranging from 10,000 amps to 65,000 amps. However, for a single-family dwelling, the available fault current rarely exceeds 10,000 amps.

Single-pole breakers

Single-pole breakers control current on loads that use only one leg of the incoming 240 volts. Standard single-pole break-

ers come in 15-, 20-, 25-, 30-, 40-, and 50-amp ratings. Most people rarely use anything other than 15 and 20. Single-pole breakers are also available in three varieties: half size, full size, and dual. The half-size breaker is as thin as a pancake and hard to turn on and off due to its tiny handle. Both full-size and dual breakers are 1 in. thick; however, dual breakers control two circuits.

ABOVE CODE Although all these breakers work in basically the same way, I generally use full-size breakers. Full-size breakers are easier to turn on and off and absorb heat better than smaller breakers do. They also fit in any panel; half-size and dual breakers fit only in panels specially designed for them. In addition, installing too many half-size and dual breakers makes it much too easy to get into "panel fill" violations, and the panel invariably winds up looking like a bird's nest with no room to work.

Compare the different types of single-pole breakers. From left to right: a common 1-in.-wide breaker, a dual breaker that only fits onto slotted buses (also 1 in. wide), and a half-size breaker. Not all manufacturers make half-size breakers, which only fit into panels designed for them.

Two breakers are inserted onto a panel bus. The dual breaker (left) cannot slide down onto a solid tab like the common single-pole breaker (right) can. A metal band inside the back of the dual breaker prevents it from being inserted into a panel for which it is not designed.

Panel-Fill Violations

A panel-fill violation occurs when dual breakers are installed in solid-tab panels; the panel then houses more breakers than it was designed to hold. To prevent this from happening, full-size panels have solid tabs that allow only full-size breakers. Dual breakers have a metal clip on the back that prevents them from fitting into a solid tab.

When a cost-conscious contractor installs a cheap, small-volume panel that holds only a few breakers, the electrician may eventually run out of room. To get more breakers into the panel, some electricians remove the metal clip so that the dual breaker can fit on a solid tab. This is not only illegal, it is also danger-ous...stuffing a panel like that can create a fire hazard.

A former electrical inspector for Roanoke County, Virginia, told me of an inspection where clips from dual breakers were littered all over the floor in front of the service panel. Looking into the panel, he saw an overabundance of dual breakers. Needless to say, the panel failed inspection. The bottom line: Install a full-size 40/40 panel. The cost difference is insignificant.

Double-pole breakers

Double-pole breakers, rated at 120/240 volts, control current on loads that use both legs of the available 240 volts at the panel. They normally consist of two single-pole breakers with one handle and a shared internal trip mechanism. To install them, simply hook the side that faces the gutter under the panel notch and push the other end onto the panel bus (the same way you install single-pole breakers). Double-pole break-ers are available in sizes ranging from 15, 20, and 25 amps all the way up to 200 amps. The 15- and 20-amp breakers are normally used for low-wattage water heaters, baseboard heaters, and pumps; 30 amps are used for water heaters, dryers, and heat pumps; 40- and 50-amp breakers are used for stoves; and breakers larger than 50 amps are used for electric heating systems.

Like single-pole breakers, double-pole breakers come in two-in-one units, called a quad. When there are only two slots available in a panel for two double-pole circuits, you may have to use a quad, but only do so if the panel has slotted tabs.

A 2-in.-wide double-pole breaker stabs onto both hot tabs of the panel's bus to power 240-volt and 120/240-volt loads. It consists of two single-pole breakers sharing a common trip.

Be careful when switching quad breakers on and off. The two inner breakers are tied together to power one double-pole device. The two outer breakers are tied together to power the second device. The long handle connecting the two outer breakers can be tricky to operate. To turn the device on, push up on the center of the handle; to turn it off, pull back on the center of the handle. However, if you force one side of the breaker up by pushing on one end of the handle, you could wind up having one side of the 240-volt circuit on and the other side off.

This quad breaker comes as a dual double pole. There is one 30-amp double pole on the two inside terminals. The two outside terminals are ganged together to make a second 30-amp double pole.

This quad breaker has a 30-amp double pole and two 20-amp single poles.

Installing and Wiring a Circuit Breaker

Hooking up a circuit branch to a breaker is fairly simple and basically the same for both copper and aluminum wire. First, strip off just enough insulation to slip the wire into the breaker. Don't strip off too much, though—if more than ¼ in. of bare wire is showing outside the circuit breaker, the wire may short out on another wire. Make sure the screw is on tight and install only one wire per breaker. (The Square D QO® line breaker is the exception; it can have two wires on its 15-, 20-, and 30-amp breakers.)

When connecting aluminum wire to a breaker, follow these instructions. Without nicking or cutting the wire, completely strip the insulation from the end. Wire-brush the exposed conductor strands to remove

any oxidation (the wire should be shiny). Apply a coat of an approved antioxidant and insert the conductor into the breaker, making sure that all strands are contained. Then tighten the screw to the specified torque indicated on the breaker. Although you may be tempted, never trim a heavy-gauge wire to reduce its size to fit into a breaker. Instead, splice the wire to a smaller-gauge wire that fits into the breaker and under the screws.

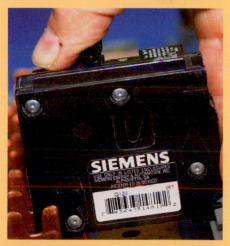

Different manufacturers attach their breakers to the panel in different ways, so it's important to follow the maker's directions. The most common installation method is to first hook the wiring side of the breaker under a panel lip, or notch. Then push the breaker down onto the bus tab—do not touch the hot bus! You should feel a solid stopping point as the breaker seats.

Limitations of Fuses and Circuit Breakers

Just because a wiring system has overcurrent protection (whether from fuses or circuit breakers), the system is not completely safe to people and appliances. Overcurrent protection is supposed to protect wiring and appliances from shorts and surges, but sometimes fuses and breakers provide no protection at all. In general, fuses and breakers do not protect against shocks, overheating, low-level faults, surges, and fires within an appliance.

Electric Shock

Overcurrent devices protect circuit wiring from overheating due to excessive current, but most do not protect you from a minor shock or even electrocution.

Overheating

If current flow through a device is impeded by a large amount of resistance, such as a loose connection, it can create significant heat. Overheating can also result from excessive current flowing through a device. Such conditions can completely burn or melt the wire insulation at the point of connection without tripping the overcurrent device.

I once did a service call on a receptacle that was not just overheating…it was glowing cherry red. It turned out that there was a loose connection behind the receptacle. A loose splice or a corroded terminal could have created the same situation. On another service call, I caused a small explosion when I let air into an arcing, corroded splice box.

Low-Level Faults

A low-level fault occurs when a hot wire makes a connection to the neutral or ground, creating a low current flow that is not great enough to trip a breaker or blow a fuse. If this occurs, it could start a fire or electrocute someone from the neutral or ground. This can happen on both line and extension cords that crack and fray, especially on those that overhang baseboard heaters. They arc and burn or they create a semiconductor bridge, allowing current to flow to a noncurrent-carrying conductor.

One time I was working on a baseboard heater in a local church. When I checked the current on both legs of the 240-volt circuit with a clamp-on ammeter, I noticed that one was pulling several amps less than the other was.

When I placed the ammeter around the grounding wire, I found the missing current—6 amps were flowing on the grounding wire. The unit was grounded, but because of an internal fault in the heater, simply touching the metal of the baseboard could have resulted in electrocution. The solution: Replace the heater.

Surges

Surges are high-voltage current pulses generated from both appliances and lightning. Fuses and circuit breakers cannot protect against these surges because the pulses only last for a few microseconds and the overcurrent protection simply cannot act quickly enough to stop them.

Surges created by appliances don't pose a significant threat to people, but they can damage electronic equipment. Surges generated by lightning can cause significant damage to both property and people, and lightning rods won't stop them. Voltage surges generated by lightning can enter a house through the power lines and then into the house through the service wires, or through saturated induction. Surges through the power lines can be stopped with surge arresters installed at the main panel (see p. 230). But saturated induction is another story. Lightning produces massive amounts of magnetic waves that simply saturate the house and the surrounding area. Anything that is a conductor will have a pulse induced within it. These pulses are created on both sides of the overcurrent protection and even in the grounding system itself. If you're touching a microwave and your bare feet are on a concrete floor, you could get a shock. If the grounding system is very good, the shock will feel like a needle jab—the same as a GFCI jolt. But if the grounding system is bad (due to high resistance), the shock could be fatal.

Fires from within an Appliance

If your appliance malfunctions internally and starts to burn, the overcurrent protection probably will not trip, which is why many houses burn down. I've seen a timer in a washing machine catch fire, and I've seen televisions, radios, and computers go up in smoke. In none of these cases did the breaker trip. Until the insulation melts and a ground fault or short occurs, there is not enough current to trip the breaker.

CHAPTER 8

Ground-Fault Circuit Interrupters

AS A PROFESSIONAL, I've used several tools over the years that have malfunctioned and placed me into the electrical path. One morning, I was drilling some treated posts outside so that I could attach electrical boxes to them. Dew was on the ground, my shoes were wet, and I was using an old, metal-case drill. While drilling the first post, the hot wire in the drill shorted over to the metal case and into the handle. Current started flowing into my hand, contracting the muscles and making my grip tighter on the tool that I desperately needed to release. The current flow went through my arm, my body, and my heart, then finally through my legs and into the ground. Luckily, the drill was plugged into an outdoor receptacle that was protected by a ground-fault circuit interrupter (GFCI). So instead of being killed, I felt just a sharp jab, like a pinprick, and then the GFCI opened the circuit, stopping the flow of current. I owe my life to the person who developed the GFCI. (Thank you, whoever you are.) Thousands of people like me owe their lives to GFCIs.

The guts of a GFCI removed from a receptacle. The black wire connects to the brass screw on the left, the white wire to the silver screw on the right. The black disk compares the current leaving from and returning to the main panel and breaks the circuit if the currents aren't equal.

GFCIs are inexpensive electrical devices. They are easy to install and available as either receptacles or circuit breakers. Just as overcurrent devices protect circuits, GFCIs protect lives. The old movie scene of a radio dropping into the tub and electrocuting the ill-fated bather could not happen in a bathroom properly wired with GFCI protection. This chapter will illustrate how GFCIs work, how to test them, and more important, where and how to install them properly.

How a GFCI Works

General-purpose, 120-volt household circuits have current flowing to and from the load on two insulated wires. Power is brought to the load on the black wire, flows through the load, and then returns via the white wire. A GFCI compares the current flowing to the load with the current coming from the load. The current,

or amperage, should be equal. If there's a difference, the electrons must be flowing somewhere other than the load (such as through you to ground), and the GFCI will open the circuit. Such a current can be as low as .006 amp and doesn't require a grounding wire to work.

The GFCI opens the circuit in $\frac{1}{25}$ to $\frac{1}{30}$ of a second, which is 25 to 30 times faster than a heartbeat. You'll still receive a mild shock, which will feel like a pinprick, but don't panic. If you receive a shock while on a ladder, freeze. Jerking your arms or legs wildly could make you lose your balance and fall off the ladder, causing more harm than the shock itself.

As good as GFCIs are, they can be fooled. As long as the current flowing through the black and white wires is equal, the GFCI operates as though everything were okay. However, if you're standing on an insulator or a nonconductive surface, such as a dry board, and you place your body between the black and white wires, the electricity will flow out

How a GFCI Receptacle Works

As long as the current on the black wire remains equal to the current on the white (return current), the GFCI operates like a normal receptacle. But if the return current becomes less than the input current, the GFCI will immediately shut off the power to the load.

Circuit Is OK

Current from the main panel is 1.5 amps.

GFCI receptacle

Appliance

Return current along the neutral is 1.5 amps.

The current on both the black and the white wires is equal.

Ground Fault Occurs

The current is missing because a ground fault has occurred. The difference current of .5 amp trips the GFCI, cutting off the power to the load.

1.5 amps

1.0 amp

The appliance falls into a pool or bathtub.

.5 amps

A GFCI Can be Fooled

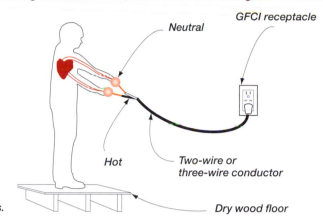

Hot

Neutral

GFCI receptacle

Two-wire or three-wire conductor

Ground-Fault Situation

The current flows from a hot wire into your hand, down through your body, and into the ground. The GFCI senses a different current and trips.

No Ground Fault

The current flows from the black wire through your body and out of your other hand into the neutral wire, with no current flowing into the ground. Since no ground fault occurs, and no current is missing, the GFCI will not trip.

Neutral

GFCI receptacle

Hot

Two-wire or three-wire conductor

Dry wood floor

the black wire, through your body, and back into the white wire. No current will leak through you to ground, so a ground fault won't exist. And because a GFCI can't tell the difference between you and a toaster, your heart will be sent into ventricular fibrillation, a wildly erratic rhythm, and you'll be dead in a few minutes. It would be as though there were no GFCI protection on the circuit at all.

In that situation, a circuit breaker would not have helped. Breakers for general-purpose receptacles trip only when the current exceeds 15 or 20 amps—2,500 times more than it takes to produce death. Breakers, and even fuses, are designed to protect the wiring within the household against excessive current. They are not designed to protect your life.

GFCIs can be used in a few unusual situations. For instance, they can be used ungrounded if there is no ground on the circuit (the power is coming from an ungrounded receptacle). The logic is that

Ungrounded but GFCI Protected

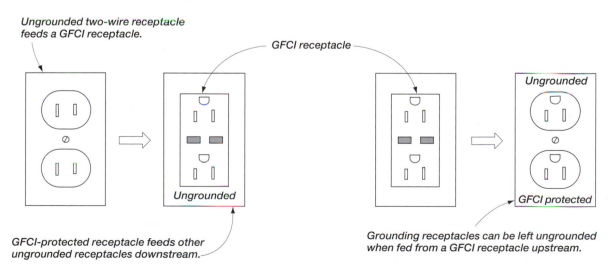

Ungrounded two-wire receptacle feeds a GFCI receptacle.

GFCI receptacle

Ungrounded

Ungrounded

GFCI protected

GFCI-protected receptacle feeds other ungrounded receptacles downstream.

Grounding receptacles can be left ungrounded when fed from a GFCI receptacle upstream.

you are safer with an ungrounded GFCI than with an ungrounded receptacle. Thus, you can replace any ungrounded receptacle with a GFCI receptacle. Just be sure to mark the GFCI as being ungrounded. In this case, it's important to note that the GFCI may protect you from a ground fault, but it will not protect your three-prong appliance in the event of an electrical surge or spike.

A GFCI can also feed ungrounded two- or three-prong receptacles. Those downstream plugs will have the same protection. The GFCI will trip if there's a ground fault in any receptacle in the circuit. As before, if a three-prong receptacle is ungrounded, you must label it as such.

Types of GFCIs

The two most common types of GFCIs used around the home are the receptacle and circuit-breaker types. Receptacle-type GFCIs, or GFCI receptacles, fit into a standard receptacle box and will protect any receptacle wired downstream. Circuit-breaker types, or GFCI breakers,

fit into the main control panel just like a standard breaker and, if wired correctly, will protect all the receptacles on the circuit.

GFCI circuits are more expensive and less convenient than GFCI receptacles. GFCI receptacles can be reset at the point of use; GFCI breakers require you to go to the main panel. However, I've found that GFCI breakers tend to last a lot longer than GFCI receptacles, and they are a good choice if you forget to regularly test GFCIs to make sure they're working properly.

GFCI receptacles are available in 15- and 20-amp configurations and both 120 and 240 volts. GFCI breakers are available in configurations up to 30 amps from most manufacturers, and a few manufacturers make them in configurations up to 60 amps. Breakers rated above 60 amps (which are used for spas) are hard to find, but they can be ordered. GFCIs are sized just like regular receptacles and breakers. For instance, a 20-amp GFCI circuit breaker goes with a circuit containing 12-gauge wire. Do not install a 20-amp unit on a branch circuit that uses 14-gauge wire.

GFCIs come in many guises. From left to right: GFCI receptacles, 15 and 20 amp; GFCI circuit breakers, 15 and 20 amp; and a GFCI built into a short multi-outlet extension cord.

Where Ground-Fault Protection Is Required

GFCIs are very sensitive—they will trip with a difference in current as low as 0.006 amp—so they should not be installed where life or property will be in jeopardy should the power be cut off. For example, freezers, refrigerators, sump pumps, and medical equipment should not be on GFCI-protected circuits. In addition, lights—unless you have a good reason to do so—should never be on a GFCI-protected circuit. If the GFCI trips, you might be left in the dark, which could be dangerous, especially if you're trying to find your way out of a wet, unlit bathroom. Also, GFCIs cannot be used for ranges, ovens, cooking appliances, and clothes dryers whose grounded neutrals are connected to the frame of the appliance.

For life protection, the NEC requires GFCIs in a few locations in a house: kitchens, bathrooms, garages, crawl spaces, unfinished basements, decks, porches, and outbuildings (and don't forget the job site, as mentioned earlier). In wet locations, such as on kitchen countertops, all receptacles must be GFCI protected. Most appliance cords are 6 ft. long, so any appliance within 6 ft. of a sink can fall into the water. Because bathrooms are almost always wet, GFCI protection is required for all receptacles, whether they are located in a light, in a medicine cabinet, or in a wall. The overhead bathroom fan or light—or combination fan/light—needs GFCI protection when it's mounted in a bath or shower enclosure. In addition, the unit must be approved for wet locations.

GFCI protection is required in any general-purpose receptacle in a garage; in a crawl space at or below grade level; in an unfinished basement; in a boathouse, dock, or seawall; and in areas around swimming pools, spas, and hot tubs. When it comes to GFCI protection, swimming pools, spas, and hot tubs, because of the obvious danger involved, are complicated and should not be handled by an amateur. Those areas should be looked at and wired by a licensed master electrician.

All outside receptacles, including those in outbuildings, need GFCI protection. If a load will be plugged into the receptacle day and night, rain or shine, the outlet must be covered. Typical examples are Christmas-tree lights, electric fence chargers, and low-voltage transformers. The receptacle cover must be suitable for wet locations while the appliance is in use. Such covers sometimes look like bubbles on the front of receptacles. You cannot use a cover with a metal, spring-latched door. That type of door is used simply to cover the outlet when it's not in use. (For more on watertight covers, see chapter 9.)

ABOVE CODE Even when exterior receptacles are covered, moisture can still creep in and destroy the mechanism that makes the GFCI work properly. The receptacle will still provide power, but the ground-fault (a.k.a. life) protection will be lost. To eliminate this threat, I recommend using circuit-breaker GFCIs for permanent outdoor installations.

Any receptacle, whether two-wire or three-wire, that goes bad and is located in an area currently requiring GFCI protection must be replaced with a GFCI receptacle. For example, if you're changing an old receptacle in a bathroom or along a kitchen countertop, the replacement must be GFCI protected.

Exceptions

There are a few exceptions to the rules. For example, if a corded appliance in a

Wiring a GFCI Receptacle

A GFCI receptacle can provide ground-fault protection to regular receptacles downstream, as long as the wiring isn't reversed. Wires from the main panel should be attached to the line side; wires feeding receptacles downstream should be attached to the load side.

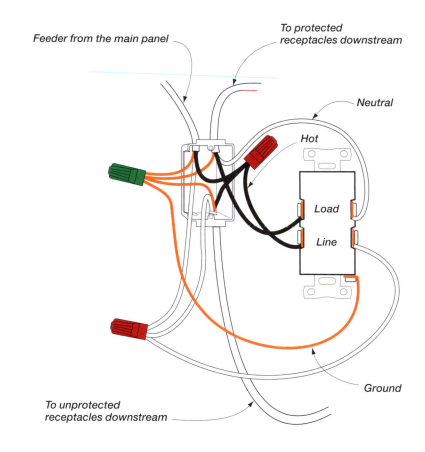

garage or basement will always be in that same location—such as a freezer or refrigerator—you don't want to worry about a GFCI tripping often and losing power. In that case, you may install a dedicated single non-GFCI-protected receptacle for the appliance (or a duplex outlet if there are two appliances) so that only that appliance can be plugged into it. Another exception is for receptacles that are not readily accessible, such as a garage-door opener.

Wiring GFCI Receptacles

In order to save lives, GFCI receptacles and circuit breakers must be wired correctly. And with so many wires involved, it's easy to connect one incorrectly. GFCIs have both load and line connections—four screws and wires to figure out, plus the ground wire. Sometimes the load connections are on the right, sometimes on the left, and other times at the top or the bottom. There is no way to tell any of the connections apart without reading the print on the back of the GFCI. And the print can sometimes be difficult to read.

Some GFCI receptacles provide power even if they are wired backward (power from the service panel is wired to the load side instead of to the line side). If the downstream receptacles are wired from the line side of the GFCI, they may not work correctly: The GFCI may not be able to sense a current imbalance. Worse, if the GFCI is wired incorrectly, it may still test normally. (Some GFCIs now have built-in sensors that light up if they are wired improperly.) To wire a GFCI receptacle, follow these steps:

1. Locate the terminals for LINE HOT and LINE NEUTRAL. Connect them to the feeder wires from the main panel (black to hot, white to neutral) to bring power into the GFCI. Once connected, protected power will be available at the GFCI and at the load terminals. Unprotected power is taken off the two feeder wires before they attach to the receptacle line terminals. Never put two wires under any screw; instead, use wire nuts and pigtail off the splice to the GFCI.

2. Identify the terminals for LOAD HOT and LOAD NEUTRAL (these may be covered with tape). These provide outgoing protected power, if you want it. If you have no need for any more protected receptacles in the circuit, nothing will connect to these two terminals. However, any receptacle that uses these two terminals for its feeder will have ground-fault protection provided by the GFCI receptacle. If used, all pro-

tected receptacles should be marked on the receptacle cover as "GFCI protected." Little stickers are normally provided with the GFCI receptacle for this purpose.

3. Connect the grounding wire. Though the grounding wire should always be connected if a grounding circuit exists, a GFCI doesn't need a ground wire to work. All it needs to cut off the circuit is a current imbalance from hot to neutral. For this reason, GFCIs can replace old two-wire receptacles.

Wiring GFCI Circuit Breakers

A GFCI circuit breaker protects wiring just like a standard breaker; that is, by tripping when the current exceeds the breaker's rating. It also provides GFCI protection, tripping when current is missing in the circuit. Before installing the breaker, make sure it matches the gauge of the branch's wiring: a 15-amp unit with 14-gauge wire, a 20-amp unit with 12-gauge wire.

Wiring a GFCI breaker is fairly simple but still requires care, because the wiring can easily be reversed. A standard GFCI breaker has a white wire attached to it. This wire, normally coiled like a pig's tail, connects to the neutral bus and is the breaker's 0-volt reference. The breaker also has another solid white wire, which connects to the neutral from the load. The black (hot) wire connects the hot terminal on the breaker to the load.

However, some breakers simply have screw terminals for the neutral and hot wires. If that is the case, take a close look at the two screw terminals—the writing is hard to read. One reads neutral and receives the circuit's white wire; the other reads hot and receives the circuit's hot wire. Do not reverse these two wires! If reversed, the circuit may test normally using the GFCI's test button. It will even trip using GFCI push-button testers on the downstream receptacles, but it may not provide ground-fault protection. The only way you will know there's a problem is by using a plug-in tester on one of the downstream receptacles to verify polarity.

Testing GFCIs

All GFCIs should be tested at least once per month with their own test buttons. The test button places a current imbalance on the circuit to see if the GFCI will open the circuit. When a test button opens a circuit, a GFCI receptacle opens both the neutral and the line terminals; it works, even when the hot and neutral wires are reversed. A GFCI breaker differs in that when it trips, it opens only the hot line—the neutral stays intact. If a breaker has been wired backward, the intact line is now the hot line and current can still go to the load—and to you. After a test, press the reset button on the receptacle to ready

Commonly used testers, both low tech (right) and high tech (older model Sure Test). Neither one should be used to fail a GFCI. The built-in test button within the manufacturer's GFCI is the only "official" way to pass or fail any GFCI.

TAKE NOTE

GFCIs are not toys. I mention this because one of my customers had a child who loved to press the test and reset buttons on a GFCI receptacle I had installed. She liked to hear the click as the GFCI tripped off. She kept doing this time after time, day after day, click after click, until finally the button broke off and flew across the room.

the GFCI. To reset a GFCI breaker, turn the breaker to full off and then on again.

GFCI testers

Many people used to (or still) rely on plug-in testers with a push button to test GFCI receptacles and breakers. However, these inexpensive testers aren't always accurate. For example, there are times when the test button of a GFCI receptacle indicates that the receptacle is working properly, but a cheap hardware store tester cannot get the receptacle to trip. In that case, the fault may lie with the tester, not the receptacle. The same logic goes for testing GFCI circuit breakers.

It's important to realize that the testing circuit built into GFCI outlets and breakers are UL approved. Most cheap testers are not. If you fail a GFCI device using a typical cheap tester, you are, in effect, saying that a non-UL approved testing device made by a third-party manufacturer can overrule a UL approved testing circuit made by that manufacturer. Do not put yourself in this position. Thousands of GFCI devices have been sent back to their manufacturers because of wrong diagnoses made by unsuitable testers.

So what should you do? When you push the test button on a GFCI receptacle, verify that power is removed (check this with a multimeter, not a light-up circuit tester). Next, press the reset button on the receptacle (or throw the breaker switch off and then back on) and check that full voltage returns. In actual field experience, I have never seen a ground-fault breaker that, when the test button is pushed, continues to give voltage. On the other hand, it is common for a faulty GFCI receptacle to continue giving voltage even after the test button is pressed, especially if the device was installed outside. Be advised that the only way to verify full voltage is by using a multimeter, not a plug-in tester. If the voltage comes back to only 70 volts due to a faulty device (internal corrosion), the plug-in tester's lights will still glow as though it had full voltage. The reverse is also true: It takes around 53 volts to light those lamps. The checker could indicate "no voltage" even though there is still plenty of current to zap you.

If there isn't a test button or if you are testing receptacles at the distant end of a GFCI-breaker protected circuit, you may need to use a plug-in tester. However, if the GFCI breaker doesn't trip, don't automatically condemn the breaker. Retest the test button on the breaker itself. If it doesn't trip, then it's the breaker's fault; if it does trip, then the fault may be due to a wiring problem with that specific receptacle.

You may also use plug-in testers to test receptacles working off the load side of a GFCI receptacle. If the test button on the GFCI receptacle works but you cannot trip a distant receptacle, write it up as a potential wiring problem, but realize that the problem may lie with your tester. Low-cost checkers use the incoming voltage as a reference. If the voltage is low, the checker may be unable to leak enough current to trip the circuit. Better-quality plug-in checkers use a device that compensates for the lower voltage.

The Final Word on Testing GFCIs

On Feb. 16, 1994, OSHA issued this statement to its field personnel indicating how to properly test GFCIs: "GFCI testers which are not an integral part of the GFCI shall not be used as a means of determining compliance...because such testers often do not produce accurate results. Only the circuit testing which is manufactured as part of the GFCI shall be used to determine whether or not the device will function as intended."

This resulted in Technical Note 91-1, titled "How to Test the Operation of a GFCI," which goes into further detail. It also describes the problems in using even the fancier, more sophisticated testers.

Don't confuse a GFCI with an AFCI. Both have white pigtails and a test button, and they wire identically into the circuit. Other than perhaps a different colored test button, the only way to tell the two apart is to read the fine print.

Arc-Fault Circuit Breakers

An arc fault is an unintentional electrical discharge characterized by low and erratic current, which can start a fire. There are three common types of arc that may occur in your home:

- Hot to neutral: The most common example is a too-tight NM connector and a too-tight staple.

- Hot to ground: An arc between a wire and ground that isn't sustained long enough to open the circuit of a common breaker.
- Series: A wire that is cut coming back in contact with itself. Extension and line cords are primary examples.

Hard to tell apart! Both the GFCI (right) and the AFCI (left) have their pigtails trimmed and attached to the neutral/grounding bus of the service panel. They both take in branch circuit black (hot) and branch circuit white (neutral) conductors.

Regular circuit breakers don't always trip in these instances because it's not something they were designed to do. For example, most breakers rely on a certain sustained amount of heat in order to trip. If the surge is short enough, a tremendous current can flow through the overcurrent device (breaker) without tripping it. (This is why lightning surges of 10,000+ amps can flow through a breaker.) Arcs that are less than the breaker's overcurrent rating also occur. For example, an arc current of less than 20 amps may not trip the breaker, but it can start a fire.

An arc-fault breaker (arc-fault circuit interrupter, or AFCI) is an overcurrent device that opens a circuit when there's an excessive load or fault on the line or when a heavy arc occurs on the branch circuit. AFCI breakers are now required for any circuit that has an outlet in the bedroom. Besides receptacles, this also applies to circuits that power lights and smoke alarms. Both 15- and 20-amp breakers are currently available.

Although they may look similar, GFCIs and AFCIs are not the same. GFCIs are designed to sense small amounts of leakage current from the hot wire to ground (ground fault). They are not designed to sense arc faults. Warning! I do not advise installing AFCIs on circuits, even bedroom circuits, that power life-sustaining equipment. Because this goes against code you might need to get an exemption.

Wiring AFCIs

AFCIs install just like any other breaker. In fact, they look like GFCIs (right down to the pigtail wire) and can be confused with them if you don't read the fine print on the breaker's sticker. To help prevent a mix-up, Siemens manufactures AFCIs with a blue test button and GFCIs with a white test button.

It's easy to get confused about where the black and white wires terminate on the breaker. Before you insert the breaker, look at the stamped plastic on both sides to determine which wire goes where. One side says "white" or "neutral," the other side says "black" or "hot." Once you insert the breaker between other breakers, these markings disappear. To install, simply snap the breaker into the panel like any other breaker and connect its white pigtail to the grounded/neutral bus in the main service panel (just like a GFCI breaker). Connect the black wire and the white wire to their respective screws.

Arc Faults

Parallel Hot to Neutral Arc Fault
The hot wire arcs against the neutral wire through damaged insulation.

Parallel Ground Arc Fault
The hot wire arcs against the ground wire or frame.

Series Arc Fault
The damaged wire arcs against itself.

Wiring a GFCI Circuit Breaker

To load hot; normally black or red

GFCI breaker

To load neutral; normally white

Neutral bus

CHAPTER 9

Receptacles and Boxes

YOU'RE ABOUT to learn all the top-secret information about receptacles, boxes, and wiring that contractors rarely discuss. You need to know what to buy and what to avoid, what looks good and what really is good, and what works and what doesn't—before you install a receptacle, not after. In this chapter, I discuss standard 120-volt receptacles seen throughout the home. (I discuss dryer and stove receptacles in chapter 12.)

Receptacle Boxes

One problem many electricians face is not thinking ahead and having the right box for the job. When buying an outlet box, consider whether it is deep or shallow, metallic or nonmetallic, and a nail-on (also called captive or integral) or a bracket-mount type. Each one has advantages and disadvantages. Not knowing which receptacle box you need leads to many problems, and one of the most common is using a box that's too small.

Volume problems

Most receptacle boxes have one common problem: They don't have enough room or volume inside the box to contain both the receptacle and the splicing for the incoming and outgoing cables. If the box is too small and doesn't have enough volume, the wires become overcrowded, resulting in broken and shorted wires and possible damage to the box's receptacle-holding threads. To prevent this from happening, the NEC regulates the amount of wires and devices that can go into a receptacle box (see NEC Section 370-16).

The regulation is determined by assigning cubic-inch volume units to everything that can go in a box and then having a fixed limit amount each box can carry (if you stuff in too many cables, it is called a cable-fill violation). Nonmetallic boxes come in 16, 18, 20.3, and 22.5 cu. in., and the volume is stamped inside the box. To determine the volume of a metal box, check the manufacturer's catalog or use an NEC chart. (For the amount of cables that can fit into other boxes, see the chart on p. 186.)

To avoid cable fill violations, always use the deepest box you can find (3½ in. deep, or 22.5 cu. in.).

ABOVE CODE Rather than go through meticulous calculations each time you want to put a cable in a box, which few of us do anyway, use the deepest boxes available. A single-gang box (the box used almost exclusively to house receptacles) is 3½ in. (22.5 cu. in.); it is almost as deep as a common stud wall. Do not run more than three 12-gauge cables or four 14-gauge cables in this box.

There will be times when it is physically impossible to install a deep box. A typical example is when a carpenter furs out a basement wall by laying the studs flat against the concrete or block wall. This makes the wall cavity only 1½ in. deep. Since you can't use deep boxes unless you knock holes in the wall for every receptacle, you must go sideways to get the volume. Don't be tempted to use the commonly available metal handy boxes just because they fit; they don't have the volume for in-and-out cabling. In this case, use a nonmetallic 1½-in.-deep, 4-in.-square box, which has an additional front plate that's raised to hold the receptacle. The cover plate must be installed before the finished wall goes up so that the finished wall can be cut around the raised area.

Metal boxes

Once the standard of residential construction, metal boxes are now used less frequently. Nonmetallic boxes are more popular because they are nonconductive, less expensive, and faster to install. However, metal boxes do have three significant advantages over nonmetallic boxes: They are strong, have more available designs, and can often be stacked. Stacking allows the rectangular switch or receptacle box to increase in size to single, double, or triple. In addition, metal boxes are still needed where the entire box is exposed and open to damage, such as along a basement or garage block wall, or in a situation where the required design is only available in metal.

Use a 4-in.-square box for shallow, 1½-in.-deep walls. The extra width provides the volume needed to hold all the wires. The receptacle mounts to the cover plate, which in turn screws to the box.

Covering the receptacle screws with a couple layers of electrical tape prevents them from shorting onto the sides of the metal box as you install it.

Metal Boxes Are Conductive

There is always a risk of shorting the hot receptacle screws against the sides of a metal box or having the wires cut into the metal as the receptacle is pressed back into the box. This is extremely common when a GFCI replaces a smaller, standard receptacle. Many electricians cross their fingers and look the other way as they shove in the GFCI. (One of my helpers blew out an entire circuit doing this.) To prevent the screws from shorting onto the sides of the metal, always wrap a few layers of electrical tape (Scotch Super 88) around the receptacle to cover the hot and neutral screws (see the photo above).

Handy, or utility, boxes are the most misused of all the metal boxes. The common rounded-corner, 1½-in.-deep box is only allowed one 14-gauge cable with switch or receptacle—anything more is a cable-fill violation. However, I don't know of any electrician, including me, who hasn't installed one or two 12-gauge cables with receptacle in a handy box. If a handy box needs to be used—for example, to fit into a narrow space—pick one that's at least 1⅞ in. deep, or even better, 2⅛ in. deep. When volume is needed in metal boxes, the 4-in.- and 4¹¹⁄₁₆-in.-square boxes are the best to have around (normal depths are 1½ in. and 2⅛ in.).

When using a pancake box, make sure that you do not create a cable-fill violation by putting anything in it other than one 14-gauge cable. Some lights have round bases with such little depth that they won't fit over the ½-in. pancake; in that case, you must recess the box in the wall.

ABOVE CODE Unless you have a specific need that requires a metal box, stick with plastic ones. The one exception to this rule is overhead paddle fans. Although some nonmetallic boxes are approved for paddle fans, I don't trust them yet. In my opinion, there is simply no way a plastic box can be as secure as a metal one. (For more on paddle fan boxes, see chapter 11.)

Nonmetallic boxes

Nonmetallic boxes are the most common receptacle boxes used in residential construction. They are both inexpensive and nonconductive. Although materials vary, the most common ones are flexible PVC plastic, fiberglass, and thermoset. These boxes are normally used throughout the house; however, the common rectangular box, such as the integral-nail box, should not be used if the entire box has to be surface-mounted in a habitable area. In addition, the common rectangular, non-metallic box should not be used for hanging lights. Boxes designed for hanging lights are listed as such and have a larger attachment screw than standard boxes do.

Receptacle boxes differ from switch boxes in that they rarely have more than a double gang (two receptacles in a box). Switches commonly use three- and four-gang boxes for all the switching required at entryways.

The most commonly used receptacle is a single-gang, nail-on box. For this type

One too many hammer hits will warp a plastic box.

Nail-On Box with Ribs

Raised ribs on this nonmetallic nail-on box help bring the front edge of the box flush with the finished wall.

Nail-on box with ribs

Ribs

Drywall will be flush to front edge of box.

The front edge of the box must be parallel, not angled like the one shown here, with the stud extending just far enough to make it flush with the finished wall.

Fiberglass boxes are sturdy but can easily be destroyed by one misplaced hammer blow.

of box, simply nail the box to the stud, leaving just enough sticking out to be flush with the finished wall. Three challenges occur here: getting the distance out from the stud exactly right, keeping the box parallel to the stud, and keeping the box from being distorted or broken as it is nailed onto the stud. To prevent the first two, take your time. Some boxes even have raised edges or ribs to help you judge the distance out from the stud. In addition, you can use a piece of ½-in. plywood as a spacer to help get the distance right. Hold the plywood to the stud and align the edge of the box with the plywood.

To prevent breaking the box as you nail it in, simply tap the nails—don't drive them hard—as the nail head comes close to the box. The more accurate you get the box mounted on the stud, the less trouble you'll have making the receptacle fit against the finished wall. If you don't want to be bothered by constantly solving these problems, install the slightly more expensive side-mount box. This box has a bracket mounted to its side that nails to the front of the stud, placing the box at a prescribed distance (normally ⅜ in.) from the stud and keeping its face parallel to the stud. This box will not work with thin wood-panel walls.

Cut-in boxes

To install receptacle boxes in existing hollow walls, special boxes, called old-work or cut-in boxes, are used. These boxes do not mount to studs; instead, they attach directly to the finished wall, using plastic ears in front and supports (either swivel ears or wings, also called spring ears or clamps) in back. The plastic ears hold to the drywall surface and the supports in back sandwich the drywall to

Cut-In Receptacle Boxes

Cut-in boxes are used mostly in remodeling. They attach to the face of the finished wall through drywall ears on the front. The support mechanism on the back sandwiches the box to the wall.

Ears

Adjustable ear

Swivel ears hold the box to the back of the finished wall.

Clamps hold the box to the back of the finished wall.

Wings, also called metal spring ears, hold the box to the back of the finished wall.

hold the box in place. In addition, you can make any box with drywall ears work like a cut-in box, as long as you have an adapter to do so. And remember, if you're installing a cut-in box in a wall with a receptacle directly on the other side, offset the new box a bit so that the boxes are not back-to-back (usually there's not enough room in the wall cavity to do this).

Cut-in boxes allow for few errors during installation. The attachment system on the cut-in box will only work if the hole is small enough for the box's plaster ears to press onto the finished wall front. If the electrician is sloppy and cuts the box hole too big, the box will simply fall through the hole. A good electrician will cut the hole exactly to fit the box. Some manufacturers furnish templates to aid in cutting out the hole. However, with other manufacturers, you're on your own.

Cut-In Adapter

Any box with drywall ears can be used as a cut-in box with this adapter, which is simply two metal straps that fold around the box to support it from behind the finished wall.

Metal straps

Straps fold around the box.

Installing a Cut-In Box

1. Trace the template, if supplied.

2. Cut a hole with a keyhole saw.

3. Pull wires through the knockout and place the box in the wall.

4. Secure the box in the wall.

Making a Cut-In Box

Here's how you can use virtually any box, even one without drywall ears, as a cut-in box. First, drill a small hole in the finished wall adjacent to a stud and insert a wire to establish the stud's exact location. Next, open a small hole in the wall so that you can see the stud. Place the box adjacent to the stud and trace around the box. Next, cut a hole and slip in the box (don't forget to slip the cable into the box). Nail or screw the box to the stud from the inside of the box. (I prefer to use drywall nails because their thin, flat heads don't protrude much and their serrated shanks prevent them from backing out.) Predrill the box so it doesn't crack. For extra insurance, apply a dab of construction adhesive between the stud and the box. The advantage to this method is that you can pick your boxes rather than having to use the common, low-volume, premade, cut-in boxes.

If you accidentally cut a hole that's too big for the box, cut a thin cedar shim (the kind used to shim doors and windows) 2 in. to 3 in. wider than the hole. Put some glue on the shim, slip it into the hole, and pull it back against the

inside of the finished wall. Next, cut a piece of drywall the exact size needed to make the opening the correct size and glue it on top of the stick.

Weatherproof boxes and covers

Made from both plastic and cast metal, weatherproof boxes are used outside the house when the entire receptacle box will be exposed. Because of aesthetics, new construction locates most receptacle boxes for outside outlets inside the exterior walls. This also allows contractors to use less expensive standard boxes with a weatherproof cover. When adding an outlet to a finished outside wall, a weatherproof box is normally attached to the outside siding. Even though the box is weatherproof, the receptacle still needs to be protected with a weatherproof cover.

There are two types of weatherproof covers: weatherproof while the cover is

A typical spring-loaded watertight cover is no longer code for outside use. While in use, the cover is open and does not provide weather protection for the receptacle or plug.

A watertight-while-in-use lid has a bubble to protect the receptacle and plug from rain and snow while a cord is plugged in.

closed and weatherproof while the receptacle is being used. These fit onto the front of any type of box in lieu of a common cover plate. Most of us are familiar with the metal spring-loaded caps that snap down over the receptacle or GFCI outlet—these are watertight only when the cap is closed and are no longer code-compliant. For continuous use—such as for Christmas lights, fence charges, and low-voltage transformers—a weatherproof while-in-use cover must be installed. Code-approved covers come in two designs: a bubble type that extends off the receptacle and wall (to house the cord plug while it's inserted into the receptacle) and a type that extends into the wall to allow the cover to be flush with the exterior wall. The latter is better (considerably less ugly), but it takes a 2 ft.-by-6 ft. wall to house both the box and the recessed cover. The bubble type is best when you want to convert an existing outside outlet by just replacing the lid.

The best performing (and best looking) weatherproof covers are recessed into the wall and have a seal between the lid and the box.

The problem with some watertight-while-in-use designs is that they are not watertight as advertised—although water does not rain in, it can leak in. To make it really watertight during a driving rain, there must be a seal around the sides and top. Another problem I've found is that the bubble does not extend far enough away from the receptacle to plug in a thick extension cord and close the lid. If the lid is not completely closed, the whole design is void. Be aware that most designs only work for light-duty, easily bendable cords.

Receptacles

In residential construction, cheap, low-grade receptacles are typically used for both light- and heavy-duty applications. Because of the poor quality of low-grade receptacles, I don't use them for even light-duty applications. They can break apart, cause open circuits, or even worse, create a high-resistance loose connection that results in a fire. While writing this chapter, I was called out on a job where the customer had a burned-out receptacle. He first smelled something burning (the insulation on the wires) and then pulled the appliance away from the wall to see the receptacle glowing cherry red. If he were not home, his house probably would have burned to the ground. His problem was a cheap receptacle and the way it was wired. Instead of worrying about color and style, homeowners and electricians should worry more about quality.

Grades

There are several grades of receptacles. Residential, commercial, hard-use, specification, federal-spec, industrial, and hospital grade are just some of the names

Good ($2 to $3), better ($5 to $6), best ($13 to $15).

Low-cost receptacles such as these are made from easily breakable plastic, are riveted together, and have little metal inside.

manufacturers give their receptacles to differentiate their use and construction. The middle grade is fine for light-use locations, such as in a bedroom or hallway. For frequent use, such as in a kitchen, garage, or shop, a heavy-duty receptacle should be used.

ABOVE CODE   For Above Code installations, I won't install any receptacle unless it is heavy-duty construction (both inside and out). Expect to pay at least $2 per receptacle instead of 89 cents. You get what you pay for. If you've ever wondered why those appliance plugs keep falling out of your receptacles, now you know.

Duplex or single

A duplex receptacle has two outlets to plug into. A single receptacle has only one. There are many instances when you may want or even need to use a single receptacle as opposed to a duplex. For example, you may want to put your freezer or refrigerator in an unfinished basement, where receptacles are supposed to be protected by GFCIs, but you'd rather not have these appliances hooked up to GFCI receptacles because of possible false tripping. A cord-and-plug appliance, such as a freezer, does not have to use a GFCI receptacle in an unfinished basement if it's connected to a single, not duplex, receptacle. It is also acceptable to use a non-GFCI duplex receptacle to power two stationary appliances.

A duplex receptacle has a tab on both hot and neutral sides. When half the receptacle will be switched and the other half will be permanently hot, break off the tab to isolate the top plug-in from the bottom one. Insert a screwdriver into the slotted tab and wiggle it back and forth until the tab breaks off. The tab is normally broken only on the hot side, allowing the neutral to be shared. When a single outlet is required in a certain situation (such as for a refrigerator in a basement) but the outlet is unavailable, some inspectors will allow you to break the tabs off both sides of a duplex receptacle to deaden the unused half. This is sometimes allowed because the intent of the code, as opposed to the written word, is being followed.

Current ratings

Receptacles are rated for the amount of current allowed to flow through them. Standard duplex receptacles used throughout a residence are rated for 15 amps—even those that are connected into a standard 20-amp branch circuit. A 20-amp receptacle looks different from a 15-amp receptacle. The 20-amp receptacle has an additional slot going sideways off the neutral slot (the wide one) to allow a 20-amp plug with a sideways prong to fit into it (see the photo at right). Both a 15-amp and a 20-amp plug will fit into a 20-amp receptacle, but a 20-amp plug will not fit into a 15-amp receptacle. The receptacle is designed this way so that you don't put a 20-amp load on a 15-amp circuit, which will overheat the wires and trip the breaker. You are required to use the more expensive

A single receptacle has only one place to plug in a load. Note the horizontal slot, which indicates that it is a 20-amp receptacle and capable of taking cords from heavy-duty, 20-amp appliances.

Breaking the tab on the "brass screw" side of a duplex receptacle isolates the top and bottom outlets for a "one switched and one not" combo. The neutral tab is normally left intact.

Safe Aluminum Wiring

Aluminum wiring that connects onto receptacles presents special problems. Because aluminum and copper have different characteristics, no receptacle may accept aluminum wiring unless it is designated as AL/CU or CO/ALR, which means the receptacle is UL approved for both aluminum and copper wiring. Leviton manufactures such receptacles. If you don't have a receptacle that accepts aluminum wire, you'll have to splice a 6-in., 12-gauge copper pigtail of the appropriate insulation color onto the aluminum wire in the receptacle box. Be careful not to put a severe bend on the aluminum wire, which breaks easily, and be sure to use an approved connector. Then terminate the copper wire onto the receptacle screws.

20-amp receptacle when there is only a single receptacle in the entire 12-gauge, 20-amp circuit.

A significant safety problem can occur when a homeowner purchases an appliance and discovers the appliance's 20-amp plug doesn't fit the 15-amp outlet. The appliance salesman may tell you that all you have to do is change the outlet from a 15 amp to a 20 amp so that the plug will fit. Don't believe him! Replacing the receptacle is not enough.

The only way to deal with this situation is to install a new circuit for the appliance using 12-gauge cable and a 20-amp appliance receptacle. (Just hope there is still room in the panel for another circuit!)

Polarity

Plugs and receptacles are polarized when they fit together in only one way. The neutral blade of the plug (the wide one) must fit into the neutral (wide) slot of the receptacle and the hot (narrow) blade of the plug must fit into the hot (narrow) slot of the receptacle. Making the neutral blade and slot wider than the hot blade and slot ensures that the hot and

Polarity of Receptacles and Plugs

In an electrical circuit, neutral and hot should be kept separate, which is called polarization. For a plug to be properly polarized, its neutral and hot wires must plug into the neutral and hot slots, respectively, of the receptacle.

Modern Style: Grounded and Polarized

A grounding pin forces proper polarization.

The wide slot is neutral.

The narrow slot is hot.

The wide slot is neutral.

The narrow slot is hot.

Older Style: Polarized but Ungrounded
Only a wide blade will fit into a wide receptacle slot, making it impossible to reverse hot and neutral.

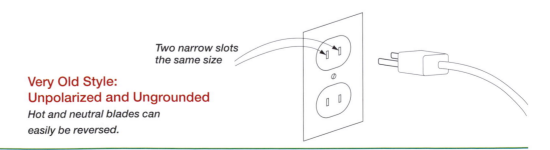

Very Old Style: Unpolarized and Ungrounded
Hot and neutral blades can easily be reversed.

Two narrow slots the same size

Cheater plugs allow you to plug in an appliance but don't ground anything unless there is a metal box and metal conduit leading back to the service panel. If the screw attachment isn't connected to the cover plate, it will pull free and expose the hot prongs of the plug, as shown.

neutral wires cannot be reversed for an appliance whose plug doesn't have a grounding pin.

Plugs and receptacles with grounding pins are automatically polarized because of the grounding pin. When the plug is inserted into the receptacle so that the grounding pin of the plug aligns with the grounding slot of the receptacle, the hot and neutral blades will slide into their corresponding receptacle slots. Some motorized appliances, such as fans and double-insulated drills, normally don't need polarization because a motor doesn't care which wire is which and the internal wiring is insulated from all metal parts. Such an appliance has an ungrounded, unpolarized plug with both blades of the same width.

Safety precautions

Some disasters are just waiting to happen. Sitting on my desk is a table lamp connected to the receptacle via an old-fashioned two-conductor cord. The plug has two narrow, unpolarized blades; it can be inserted either way into the receptacle and the lamp will work just fine. However, for safety reasons (it's also a code requirement), the center contact of the lamp shell should be connected to the hot wire of the branch circuit and the screw base of the lamp should be connected to the white neutral. But because the plug isn't polarized, you have a 50/50 chance of getting it right—or wrong. If the bulb is out of the lamp and the screw base is hot because of incorrect polarization, someone could easily be electrocuted simply by touching the female threads of the screw base. (I know of at least one child who was killed by such an occurrence.)

To ensure that this doesn't happen to you (or your kids) follow these steps:

1. Always leave a bulb in a lamp if it's plugged in.
2. Check the continuity between the center contact of the light's screw-in base and the plug blades to determine which blade is connected to the center contact. Once that's determined, put a small dot of paint on that plug's side and a corresponding dot near the receptacle's small slot. You now have correct polarization: Hot to center contact, neutral to screw base. (Of course, the safest solution is to replace the plug with a polarized one.)

Old-house polarization

Old houses with unpolarized receptacles present a unique problem. All electricians know there is no simple way to tell hot from neutral when the wiring is so faded that you can't tell black from white. Here's how to tell which is which: Run a wire from the panel neutral to one lead of a VOM and the other VOM lead to one of the receptacle slots or wires. The neutral should read 0 volts and the hot should read 120 to 125 volts. If the system is in conduit (you'll know if you have conduit by seeing it enter the panel), the conduit can be used as a ground reference (in lieu of neutral) for the test. (This is also a good way to test if the conduit can be used as a ground conductor.)

Wire Connectors

Typical wire connectors.

Twist-on wire connectors are used for small-diameter (10 gauge and smaller) copper wire splicing. These connectors are a fast, easy, and inexpensive way to cover the splice. Splices also can be crimped together, with an insulating sleeve slipped over the crimp. However, such a splice will not allow you to add or remove a wire without destroying the splice.

Most connectors are made with a thermoplastic shell over an expanding square-wire spring. The spring cuts slightly into the wire's surface and supplies the inward pressure needed for a tight connection. Shells can be round or winged. I prefer the latter because it provides better turning leverage.

Types of Connectors

More than 60 years ago, Ideal Industries, Inc. created the original Wire-Nut® (and it still owns the trademark). Today, there are several different manufacturers; their products are known as "wire connectors," not "Wire-Nuts." Electricians use four standard colors—yellow, red, blue, and orange—and each color is for a different wire combination. In general, yellow connectors can splice three 12-gauge wires or four 14-gauge wires; red connectors can splice three 10-gauge, four 12-gauge, or five 14-gauge wires; blue connectors can splice four 10-gauge, five 12-gauge, or six 14-gauge wires. I use orange connectors to splice NM cable onto the small stranded wires of light fixtures.

3M makes a connector with a long, flexible skirt. The skirt, which bends with the wire, provides better protection against bare wires sticking out of the connector. This allows for a faster turning radius when getting the wire from point to point and better wire placement within the box. 3M's connectors have wings for added turning leverage.

King created a niche in the connector market by designing wire connectors approved for wet locations—other manufacturers now make them as well. Moisture is the enemy of all electrical splices and must be kept out of the connection at all costs. I used to squirt silicone into wire connectors to keep the splices dry, but wet-location connectors are prefilled with silicone.

Silicone-filled nuts protect against sparks, water, and corrosion.

Wiring and Installation

The wiring and installation of receptacles is one of the most error-prone areas in residential wiring. The installer assumes that the ground terminals must go down, polarities are reversed, receptacles have the wires attached to them in an unsafe manner, and the wiring is installed to the receptacles in a design akin to cheap Christmas-tree lights.

People always ask me, "Does the grounding terminal go up or down?" and "Is there a bottom or top to a receptacle?" The answer is that there is no "official" right or wrong way to orient the receptacle—the NEC doesn't specify—but you can figure this out for yourself by using logic and common sense. For example, clothes washers, refrigerators, and window air conditioners typically have cords with immediate-turn plugs. For these appliances, orient the receptacle so that the plug can insert into the receptacle without having to loop over itself. Otherwise, the downward pull of the cord will tend to pull the plug out of the receptacle.

Other than for immediate-turn plugs, I prefer to put the grounding slot on top. The reason is that if a plug were partially pulled out of the receptacle, exposing the hot and neutral blades of the plug, and something metal fell on it, a direct short would not occur because the grounding pin would deflect the item away from the hot terminals.

Receptacles can also be positioned horizontally. When this situation occurs, place the grounding slot to the left and the wide neutral slot on top. That way, if a plug has its blades partially exposed and something metal (such as a picture frame) falls onto it, the object will hit the grounded neutral blade, not the hot blade.

With the grounding slot on top, a falling metal object, such as a picture frame, will bounce off a partially exposed plug with no harm done. With the grounding slot on the bottom, an object could short across the hot and neutral blades of the plug.

Ground

Preferred   **Not Preferred**

A receptacle installed horizontally should have the neutral (wide slot) on top. A falling metal object will make contact with the neutral blade of a partially exposed plug, not hot.

Neutral

Preferred

Hot

Not Preferred

Identifying screw terminals

A receptacle has three locations to accept wires:

- A hot side with two gold screws internally connected to the narrow slot of the receptacle.
- A neutral side with two silver screws internally connected to the wide slot of the receptacle.
- A green grounding screw internally connected to the receptacle's grounding slot.

The electrician connects the black or red (hot) insulated wire of the incoming cable to the gold screw, which can be traced back to the main panel's circuit breaker. The white insulated (neutral) wire is connected to the silver screw, which can be traced back to the neutral bus on the main panel. The green screw

Which Cable Is Which?

If you always bring power into a box via one particular knockout and run outgoing cables through others, you'll always be able to tell at a glance which cables are outgoing power and which one is incoming power (the author uses the upper left).

Incoming power

Outgoing power

Outgoing power

Outgoing power

Push-in terminals keep me busy trouble-shooting dead receptacle strings. Don't use them—they may become loose and pop out of the receptacle back.

is wired to the incoming ground wire, which eventually finds its way to earth ground via the main panel. The black and white wires provide a current path to and from the load plugged into the receptacle. The grounding wire (called the equipment-grounding conductor) places the frame of the load at 0 volts, or ground potential, and provides a return path for the current to the panel should a ground fault occur.

ABOVE CODE Always bring power into the box through one particular knockout—I prefer to bring power through the upper-left knockout. Doing so enables you to quickly identify the incoming cables and the outgoing cables.

What to do with extra wires

Each screw on the receptacle can have only one wire under it. If you've got power coming into the receptacle and the receptacle also feeds additional circuits, do not be tempted to twist two wires

together and tighten them both under one screw. This is dangerous because the wires will eventually work loose, resulting in an intermittent circuit—if the receptacle doesn't break apart first.

You may also be tempted to put any extra wires in the push-in terminals—available on some receptacles—or you may find it easier to attach all wires in the push-ins. Don't do it. I once had a service call in which the owners were having so many problems with the receptacles in their home that they thought the house was haunted. When I put my hand close to several of the receptacles that had nothing plugged into them, I could feel heat—enough to melt the child-protector inserts. Other outlets were generating enough heat to deform the kitchen-appliance plugs. A few hours earlier, sparks had flown out of an outlet that had nothing plugged into it.

This Hubble design wires very quickly.
Strip off 1 in. of insulation from the wire
end, then insert it from the front so that it
barely protrudes from the built-in slot.

Twist the wire around the screw, then
tighten it.

Using Pigtails

*When you've got more wires than screw
terminals, use the pigtail method to splice
wires together, then attach one pigtail to
its respective terminal on the receptacle.*

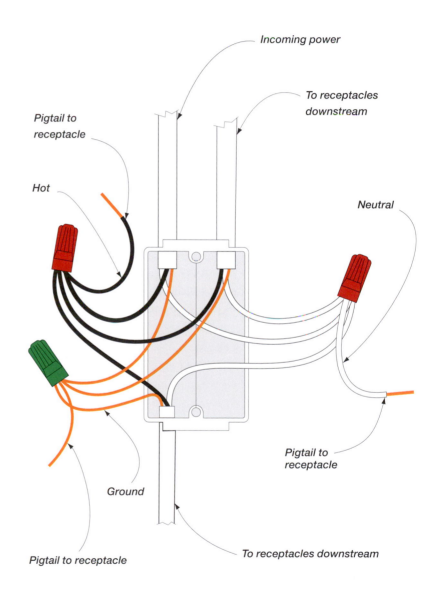

The problem was that the installer used push-in terminals, which seem as though they were designed for lazy electricians and inexperienced do-it-yourselfers. Going by various trade names such as EZ Wire®, Quickwire, and Speedwire, these systems use the spring tension of a metal "check" (similar to a clamp) inside the receptacle to hold the wire after it is pushed into the hole. They don't always clamp solidly onto the inserted wires (especially when dealing with small-diameter 14-gauge wire), causing receptacles to overheat or a total loss of power to one or all receptacles in a string. In the interest of safety, never use push-in terminals. Since January 1995, push-in terminals have been banned from 20-amp circuits. Receptacles made after that date prevent the insertion of

Making the Perfect Splice

A good electrical splice is first a good mechanical splice. If the splice is not tight, a fire can occur due to arcing. Although a standard shell doesn't support fire, a loose connection destroys the connector to the point where an arcing wire's flash can start a fire someplace else. Making a perfect splice isn't hard, but it does take some practice. Here's how to do it:

First, pull all the wires that are to be spliced together to one side of the box (for example, all black wires to the left, all white wires to the right, and all bare wires to the bottom), then cut them the same length. Make sure the wires come straight out and that they do not intertwine or loop in and out with other wires. Don't make the wires the same length by forcing some of the wire back into the box—the extra wire length takes up too much space in the box.

Once the wires are cut, strip the insulation off the wire ends. The amount of insulation you need to strip depends on the size connector you're using. It's better to strip too much (and cut some wire back) than not enough. Typically, ⅝ in. to ¾ in. should suffice.

Use electrician's pliers to twist the wires together, turning in a clockwise direction. Each wire should spiral around the others in the group. Make sure that no wire slips out and simply twists around the others. If necessary, cut the twist back ⅛ in. to even out the conductors.

Pick the appropriate connector and twist down onto the group of wires. You will know when to stop twisting when you feel significant resistance. If you don't feel much resistance, use a smaller connector. If bare conductor shows past the wire connector, use a larger one.

Splicing with a Wire Connector

1. Cut all wires the same length.

2. Strip off enough insulation that no bare wire sticks out of the connector.

Twist clockwise.

3. Twist the wires together clockwise, using broad-nosed electrician's pliers. Then insert the wire connector over the wires and twist clockwise.

12-gauge wire into the push-in terminals; however, they are still able to take 12-gauge wire under the screw terminals. In my opinion, this is too little, too late: Push-in terminals should be banned altogether.

So what do you do with the extra wires when there are no extra screws and you don't want to use push-in terminals? One method is to purchase a high-quality receptacle that accepts four hots and four neutrals using screw pressure on a metal bar to hold the wires. Otherwise, the wires must be spliced together mechanically with an insulated cap or a wire nut, with one short wire, called a pigtail, extending from the splice to connect to the receptacle (see the drawing on p. 177).

Pushing wires back in the box

Once wired, the receptacle should not just be shoved back into the box, bending and crumpling the wires in a haphazard manner. If only one receptacle and just a few wires are in the box, there shouldn't be a problem; however, if there are several wires and a couple of receptacles, there's a trick to getting all the wires back in the box without using a hydraulic jack. First, strip all the cable sheaths ¼ in. to ½ in. from where they enter the box, rather than having several extra inches taking up box space. Next, cut off all excess lengths of wire; 6 in. from where the cable enters the box is plenty. Bring all the grounds to one side, the neutrals to another, and the hots to still another; don't have wires threading through each other, making the box look like a bird's nest.

Once everything is hooked up, neatly push the wires into the box. Do the ground wires first and use side cutters to push them up against the back of the box. Next, push the neutrals to another part of the box. Push the hot wires in last, routing them along the edge, not the center, of the box. The center of the box will need extra room for the receptacles. If the receptacles don't fit after all this, you'll have to increase the size of the box or remove some cables.

Testing

Once the receptacle is wired and powered up, it should be tested with a plug-in tester to make sure it's wired correctly. The basic tester will indicate whether the wires are connected properly. A more sophisticated tester can also tell whether there is a bootleg ground (neutral shorting to ground within a few feet) and can electronically create a 15- and 20-amp load.

Receptacle strings

There are two ways of wiring strings of receptacles together: a series string,

This cutaway view shows the order in which to fold wires back into a box: first the grounds, then the neutrals, then the hot wires. The pigtails to the receptacle are prebent so that they fold neatly into the box.

■ WORK SAFE
■ WORK SMART
■ THINKING AHEAD

Don't be afraid to tape a wire connector onto a splice, but first make sure that the connector is snug onto the wires. You cannot tape a loose connector onto spliced wire ends.

such as Christmas-tree lights, and a parallel string. The former is faster and cheaper but more problematic. The latter is a better system and the one I recommend.

SERIES STRING This string is the most common. A series string has the incoming current from the panel flowing through all the receptacles of the string. The disadvantage of this wiring setup is that if a wire comes off one receptacle, every receptacle that gets power through that receptacle will also lose power (see the drawing below).

PARALLEL STRING In a parallel string, no current flows through a receptacle unless a load is plugged into it (see the drawing on the facing page). Receptacles upstream can even be removed without cutting off power to the ones downstream. To do this, the wiring must be

pigtailed. With this system, the incoming black and white wires (the ones bringing in the power) are connected via wire nuts to the outgoing black and white wires that take power to the next receptacle. This takes the power directly in and out of the receptacle box without going through the receptacle itself. To get power to the receptacle, place 6-in. pigtail wires under the receptacle screws and take them to the two splices.

ABOVE CODE Receptacles should be installed via pigtails in a parallel circuit. Wiring this way ensures that downstream receptacles stay powered regardless of what's happening upstream.

COMBINATION SYSTEM What normally happens when electricians are wiring a house is a compromise between the two systems. As long as there are not more than two black or two white wires

Receptacles in Series

These receptacles feed others downstream. If one is disconnected or removed, any others downstream will lose power, too.

The last receptacle in a string does not power any others.

Receptacles in Parallel

No current flows through a receptacle unless a load is plugged into it. Any receptacle in this string can be removed without affecting current flow to others.

Wires attach directly to the last receptacle in the string. They do not pigtail.

attached to the receptacle, the attachment is normally done under the screws (series). If there are more than two wires, they should be pigtailed (parallel).

Troubleshooting

Solving problems is the most time-consuming part of the job. A good electrician not only knows what to do when things go right but also what to do when things go wrong. Here's a list of common problems.

BOX HAS STRIPPED-OUT RECEPTACLE-HOLDING THREADS   One common installation problem is stripping the box's female threads so that the receptacle's screws cannot pull the receptacle up to the box. This normally happens when you do not bend the wires back into the box but rely on the screws to pull the receptacle into the box. (Even when you're careful, this problem eventually happens to everyone.) What do you do

about it? Drywall screws work nicely and they're easily obtainable on any job site. A drywall screw has a flat head, so the faceplate will still fit tightly against the receptacle.

BOX'S RECEPTACLE-HOLDING THREADS HAVE BROKEN AWAY
What do you do if the entire female threaded portion that holds the receptacle to the box breaks off? Obviously, you can replace the box, which is not much of a problem if the finished walls aren't up yet. But if they are up, this can be very problematic. And besides, this only happens at the end of a long, hard day when everything else has gone wrong and you have neither the time nor the patience to change a box that is already surrounded by drywall. Rather than pull your hair out, try this: Squirt a dab of construction adhesive into the area where the screw will attach. After it hardens, predrill it and use a drywall screw.

WIRES WON'T FIT INTO BOX   Another common problem is having too many wires in the box and a receptacle that doesn't seem to fit. Cable-fill violations aside, you can get more wires into a box by being neat and cutting away all excess sheath and wire.

BOX WAS INSTALLED INCORRECTLY
If the receptacle box was installed improperly, you'll have a hard time installing the receptacle. If the receptacle box extends beyond the finished wall, so will the receptacle and the faceplate, leaving an unsightly gap behind both. In this case, move the box in or grind down the front edge so that it's flush with the wall (grinding also removes female screw attachment threads for the receptacle).

If the box is set too deep within the wall cavity, you will be in code violation (the box must be flush with the finished wall) and the receptacle also will be set too deep in the wall cavity. For noncombustible surfaces such as tile, the box cannot be recessed farther than ¼ in. from the finished wall. If the drywall is cut perfectly around the box, the plastic ears (I call them Mickey Mouse ears) will affix to the drywall. But it's rare that you get a tight cut; normally, it's just the opposite. Spacers can be placed between the receptacle and the box to extend the receptacle out to the finished wall. I use small washers, nuts, and Mickey Mouse ears with long attachment screws (always

break off and save the little ears from receptacles you're throwing away; they make perfect washer-type spacers for just this purpose).

Some manufacturers make adjustable extensions for standard single-gang plastic boxes. They are hard to find, so you may have to check several stores. Beyond that, I cut the attachment nails with a reciprocating saw and pull the box forward, remounting it with drywall nails and glue. If the box is installed slightly canted to the left or right, the two horizontal slots on the receptacle (one above and one below) allow you to plumb the receptacle.

CUT WIRES   Sometimes the biggest headaches come from other people's work. On one job, the drywall installers used a broken drill bit on a router to cut out each hole for the receptacle boxes. They first attached the drywall, then plunged the bit through the drywall and into the box, cutting their way toward the box edge. Once they knew exactly where the box edge was, they inserted the router on the outside of the box and routed out the receptacle hole. This process was fast and easy (for them), but in the process, they cut every wire in every box, including the telephone wires. It was a good thing I was being paid by the hour on that job. I now tell drywall crews and the person in charge, "Whoever cuts the wires pays for it."

CHAPTER 10

Switches

SWITCHES CONTROL the power used throughout the house. For proper control, the right switch and box must be installed and the switch must be wired correctly. Installing and wiring switches is not very difficult. In this chapter, I discuss types of switches—which grades are right for different situations—and how to wire them correctly. I'll also show you how to make the job simpler by following a plan and using some common sense. It all begins with choosing the right switch box.

Switch Boxes

Because there are so many boxes on the market, the initial challenge is knowing which type of box is needed for each switching situation. Normally, a single switch requires a single box, known in the trade as a single-gang box. Two switches normally use a double-gang box, three switches require a triple-gang box, and four switches require a four-gang box, or quad. I don't recommend installing more than four switches at a single location, since nonmetallic boxes larger than four gang aren't available. In addition, switch faceplates larger than four gang are hard to find and may be expensive.

There are times when I need to install a switch in a wall in which the studs are too close to allow the correct-size box (for example, I might need a triple-gang box but only a two-gang will fit). In this situation, I use a double switch, which has two switches on the same yoke, or frame. The double switch allows me to use a smaller box—a single instead of a double, or a double instead of a triple. I also use double switches when I forget to wire in a switch or when a design is changed at the last minute. And because a double switch looks different from the other switches in the box, I can easily identify what each switch does.

Nonmetallic boxes with integral nails are the most frequently used switch boxes in residences. However, these boxes are easily damaged and so must be kept within the wall cavity. Metal boxes are rarely used within the wall cavity because of their higher cost and extra labor to install (you must ground the box). However, metal boxes are required on a surface-mount, unfinished basement or garage wall, where they take a certain amount of abuse, and conduit must be run to protect the wiring at least up to 8 ft. overhead.

Switches use the same basic boxes as receptacles but tend to be ganged more. These three are the most common: single, double, and triple.

Quad boxes need to be mounted on both sides or else they move around in the wall. Add another stud on the opposite side for extra support.

Avoid these boxes. The quad box at bottom doesn't have a support on both sides (this one has its support broken off; others are designed without them). Never use a box like the two rectangular types (1½ in. deep) above left and center, or this pancake one, above right, for anything except a single 14-gauge cable.

One important rule applies to all flush-mounted switch outlet boxes (this also applies to flush-mounted receptacle boxes): They must be installed flush or extremely close (approximately ⅟₁₆ in.) to the finished wall surface. This is to prevent fire within the room from entering the wall cavity—or vice versa. If the box is set too deep and cannot be moved, metal and plastic box extensions are available (the metal ones are extremely hard to install). If the box is set too far out from the finished wall, you must cut it off flush with the wall or somehow set it back within the wall.

Keep in mind that metal outlet covers must be grounded. This is done via the metal attachment screw. The screw grounds the cover to the switch, and the switch is grounded via the equipment-grounding conductor, AC armor, or metal conduit. If the outlets are not grounded, you cannot use a metal coverplate. In general, I don't recom-

This switch outlet box was installed flush to the stud and is now in too deep. If you can't move the box forward, install a plastic box extender, which is available from most electrical distributors.

mend metal coverplates at all. I've seen too many come loose, fall onto a plug's prongs, and create a short.

Single-gang boxes

A single-gang box with one switch is the most common switching arrangement in a residence. Even though switches normally require less volume in the box than receptacles do, you can easily put yourself in cable-fill violation if you purchase a box that's too small in volume. I never use a box smaller than 18 cu. in., and I prefer to use the bigger 20.3- and 22.5-cu.-in. boxes that I use for receptacles, even though they are significantly more expensive than 18-cu.-in. boxes. As with receptacle boxes, nonmetallic switch boxes have their volumes stamped inside; metal boxes do not.

Double-gang boxes

Switching via two switches in a double-gang box allows you to control power to two different loads from one location. Typical examples are at the main entrance, where outside lights are switched on at the same location as interior lights, at a junction of a room and a hallway, and at a box that controls both a ceiling fan and a light (see the drawing on p. 186).

A double switch can be used in a single-gang box when a double-gang box cannot fit within the framing (for example, next to a door).

■ **WORK SAFE**
■ WORK SMART
■ THINKING AHEAD

If you must cut the front off a nonmetallic box because it extends beyond the finished wall, be careful not to cut off too much, thus removing the female attachment threads.

Two Switches, Two Loads

This setup allows two loads to be controlled from one location.
Each switch is fed from one cable, but each powers a separate load.

Incoming power

Double-gang box

Hot

Ground

Neutral

To load

Number of Cables Allowed in a Single-Gang Box

Single-Gang Box Types and Sizes	Number of 12-Gauge Cables with Switch	Number of 14-Gauge Cables with Switch
Metal handy box, 1½–1⅞ in. deep (11.5–13 cu. in.)	0	1
Metal handy box, 2⅛ in. deep (16.5 cu. in.)	1	2
Metal box with integral nails, 2½ in. deep (12.5 cu. in.)	0	1
Metal box with integral nails, 2²⁷/₃₂ in. deep (15.8 cu. in.)	2	2
Metal box with integral nails, 2⅞ in. deep (18.8 cu. in.)	2	3
Nonmetallic box, 16 cu. in.	2	2
Nonmetallic box, 18 cu. in.	2	3
Nonmetallic box, 20.3 cu. in.	3*	3
Nonmetallic box, 23 cu. in.	3*	4

*Cable-fill violation, because 24.75 cu. in. volume required.

Double-gang nonmetallic switch boxes have a variety of attachment mechanisms (depending on the manufacturer); they are normally available in volumes of 25 cu. in. to 34 cu. in. Smaller boxes should be avoided because they have scarcely enough volume to accept one incoming power cable and two load cables—the bare minimum required in a box with two switches.

Triple-gang boxes

It is very common to have three switches at a house's main entrance to control the outside light, the entry light, and the interior light. Three switches (in a triple-gang box) are common whenever an overhead fan and fan light occupy two spaces in the box and a third switch is needed to control the hallway light or another load. Triple-gang nonmetallic boxes are typically available in volumes of 45 cu. in. and 54 cu. in. I recommend using the larger box—you'll be surprised how quickly it fills with cable. These boxes can be attached via integral nails or metal brackets.

Four-gang boxes

A four-gang, or quad, box with four switches is the largest switching arrangement I recommend using within a residence. If you must use a box larger than a four gang, make sure the switches are clearly marked: It's hard enough to remember what four switches do, let alone additional ones. A four-gang arrangement can be used at the main entrance to control entry lights, room lights, spotlights, and outside lights from one location. Or, if there is a hallway between the dining and living rooms, you may want switches to control a fan and a fan light, as well as lights for the dining room, living room, or hallway. The problem is remembering which switch is which. If this situation arises, you may want to pull the two fan switches away from the others and install them in a double-gang box.

Switch Grades

Switches, like receptacles, are available in many grades and vary in terms of safety and quality. The spectrum starts with a thin, low-cost, residential or general-use grade and ends with an expensive, high-quality spec grade.

The most common switch installed in a residence is residential grade, which is normally rated at 15 amps/120 volts AC (switches normally are not available in 20 amps or higher). In theory, a residential grade can have full-rated current through the switch for loads such as fluo-

rescent and incandescent lighting and inductive and resistance loads for heating. For motor loads, a residential-grade switch shouldn't be used for more than 80% of its rated current and not beyond the horsepower rating of the switch, which is normally ½ hp at 120 volts. In my opinion, a residential-grade switch shouldn't be used for anything that pulls current close to the switch's maximum rating.

Residential-grade switches are inexpensive, and you get what you pay for. It's not uncommon for this grade of switch to fall apart during installation or to die an early death if a load is pulling close to the switch's rated current. Occasionally, the side screws strip out in the body, even with just minimum pressure, and the cheap plastic body breaks apart as the screws are tightened.

ABOVE CODE Most cheap switches are designed to be wired via push-in terminals. Never, under any circumstance, use the push-in terminals—always use

Comparison of a spec-, or heavy-duty, grade (left) and a construction-grade switch (right). Some manufacturers color-code their heavy-duty switches as per different amperage specs.

the screws. I've lost track of the number of push-in connections that have pulled out of the switch and created an open circuit. Another danger with a push-in terminal is that the wire can loosen over time. A loose connection increases resistance and can lead to arcing and burning. This type of cheap push-in terminal should not be confused with the high-quality switch connections that bring in wires from the back but also have screws that tighten a bar to apply pressure on the conductors.

Spec-, commercial, and industrial-grade switches are also available for those who want something other than minimum grade. These switches have thick bodies—about twice as thick as those of residential-grade switches. They are also available in 20- and 30-amp ratings. A typical commercial-grade switch only costs $1 to $2 more than a cheap, residential-grade one. This small extra cost increases the quality tremendously. Those extra couple of dollars buy you the following features: voltage increase from 120 to 277; 20- and 30-amp designs that increase allowable current and horsepower rating from ½ hp to 2 hp; the ability to wire to side screws—no push-in terminals; a nylon toggle to resist breakage; silver cadmium contacts for longer switch use; and an internal neoprene rocker for longer switch wear.

Types of Switches

There are many types of switches, and your situation will dictate which type to use. (The most common switches in a residence are single pole, double pole, three way, and four way.) The way switches are wired has changed a lot since the early 1900s, when push-button switches were used. Today, there are many more wires going to and from switches, and an electrician must try to keep them straight. As with receptacle boxes, I normally run the incoming power cable into the switch box through the upper-left knockout and run loads out through the bottom. This way, you can tell whether the power and load conductors are wired to the switch just by glancing at it. Even if you prefer to run the incoming cable through another knockout, be consistent.

Single-pole switches

The basic on/off snap switch is referred to as a single-pole (SP) switch, because only one contact is opened or closed at any one time. It is the most common switch in a residence and is used exclusively to turn lights on and off from one location. An SP switch allows electrical current to flow through it in the ON, or "make," position and opens the connection in the OFF, or "break," position. The most common mistake in installing a simple switch is installing it upside down, making up OFF and down ON.

An SP switch has two terminals or screws: one terminal for the incoming hot wire and one for the wire called the "return," which takes the power to the load when the switch is ON. As far as making the circuit work, it makes no difference which wire connects to which terminal. But I prefer to follow a system: I always run the incoming hot wire to the top terminal and run the wire to the load off the bottom terminal. This way, I can tell at a glance which wire runs to the switch. It also is an extremely effective troubleshooting aid.

How you wire an SP switch depends on whether the power cable runs to the switch box or the load box. If a load, such as an overhead light, runs to the load box, a two-conductor cable must connect the light to the switch. One wire, the white

Two Methods of Wiring a Single-Pole Switch

White wire (taped black) becomes hot feed to the switch because white is not allowed to provide power directly to the load.

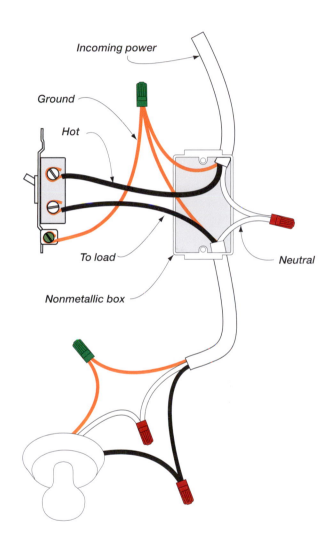

One Cable from Load

Incoming power cable (feeder) is brought directly to load, or light. Splice is made at load, not in switch box. From splice, one cable goes to switch. White wire to switch should be hot, not black wire, because black wire of same cable must be hot (switched) feed to load. Adding black tape to hot white wires to indicate they're hot is required.

Two Cables: Incoming Power and Load

Power is brought through black hot through switch, to load. White neutral wire from incoming cable is spliced directly to white from load. Ground wires from both cables are spliced together with ground from switch.

(taped black), takes the power from the light outlet box to the switch; the black wire brings the switched power back to feed the light. I don't like this system because it places the power cable in a location that's hard to access for troubleshooting.

ABOVE CODE I prefer bringing both the power cable and the load cable (from the light) into the switch outlet box. I splice the neutral from the power cable straight through to the load, take the hot wire to the switch, and then run a wire from the second switch screw to the light. It's simple, logical, and easy to troubleshoot. If you want to tap into the power to feed a receptacle, simply splice onto the hot cable as it enters the box.

Wiring a Double-Pole Switch

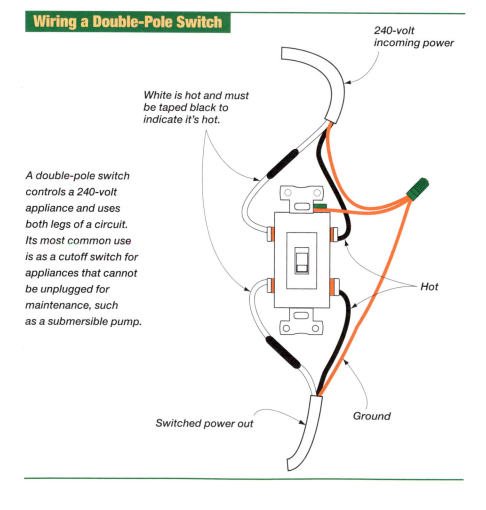

240-volt incoming power

White is hot and must be taped black to indicate it's hot.

A double-pole switch controls a 240-volt appliance and uses both legs of a circuit. Its most common use is as a cutoff switch for appliances that cannot be unplugged for maintenance, such as a submersible pump.

Hot

Switched power out

Ground

Double-pole switches

A double-pole (DP) switch can turn two independent circuits on/off at once because two contacts, or poles, are opened and closed at the same time. A DP switch is wired as two independent SP switches—one on the left and one on the right. In this situation, the white is hot, not neutral, so it must be taped black to indicate that it's hot. A DP switch is normally used to control a 240-volt appliance, such as a submersible pump or other piece of equipment without a plug that can be disconnected for maintenance. The appliance therefore must have a cutoff switch adjacent to it or its controls to protect the person performing the maintenance. I wire a DP switch the same way I wire an SP switch: hot to the top, load to the bottom (see the drawing below left).

Three-way switches

A three-way switch controls a load from two locations, such as at the top and bottom of a staircase or at each end of a hallway. A three-way switch has three terminals: a COM (common) terminal, noted by a dark-colored screw, and two terminals for the hot traveler wires that have no polarity (the switch has no ON or OFF). The COM terminal is also called "the tongue," which explains how the switch works: It laps up to connect to one traveler terminal and down to connect to the other.

Here's the simplest way to wire a three-way switching circuit (see the drawing on the facing page). Run the hot wire to the COM terminal of the first three-way switch. Run the neutral (of the cable with the hot wire) to the white load wire. Run the black load wire to the COM terminal on the second three-way switch. From the two leftover terminals of the first switch, run the two traveler wires (no polarity involved) straight to the two leftover terminals of the second switch.

Always use three-way switch cable, not standard two-conductor cable, and run the cable from the first switch to the load and then to the second switch.

Four-way switches

For additional point-of-load control between three-way switches, install a four-way switch. There is no limit to the number of four-way switches that can be installed, as long as they are installed between the three-way switches (a three-way switch must always be on each end of a multiple-point switching circuit). Four-way switching is used for rooms with three or more entries with switched lighting at each one.

The switch's box may have a schematic on how to wire the switch. It's easy to wire four-way switches incorrectly, so use a continuity tester to verify the switching arrangement (see the drawing on p. 192). I connect the incoming travelers to the bottom switch terminals and the outgoing travelers to the top terminals (verify the terminal connections). Don't attach each set of travelers to one side of the switch. Doing so will throw a short on the circuit.

Three-Way Switch Rules

Worried about whether you wired the three-way switch correctly? All three-ways, no matter how complicated, follow three rules. Get them right and the wiring is right.

1. Power cable hot wire (normally black) to one COM terminal.

2. Power cable neutral straight to the load neutral and the load hot wire to the opposite-switch COM terminal.

3. Connect the travelers (no polarity) to the two leftover screws.

Specialty switches

Switches are available for just about any switching condition that exists. Lighted toggle switches that glow when off are perfect for dark locations or even a child's room so the child always knows where the light switch is. Pilot-light switches that glow when the load is turned on are very useful when a switch is at a distant location from the load. Keyed switches are available for situations that require extra security. Timer switches are available for temporary loads, such as bathroom fans. Programmable timer switches are available to turn on lights, a radio,

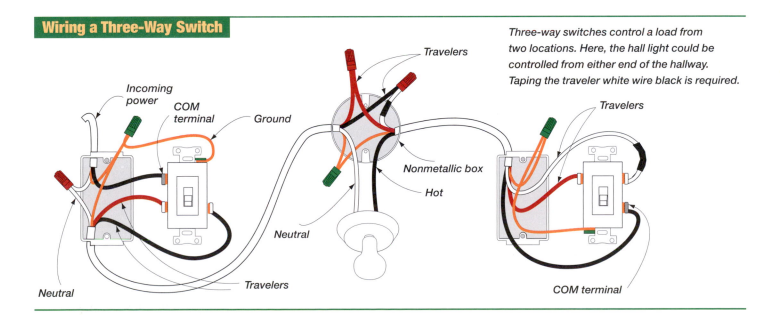

Wiring a Three-Way Switch

Three-way switches control a load from two locations. Here, the hall light could be controlled from either end of the hallway. Taping the traveler white wire black is required.

Incoming power

COM terminal

Ground

Travelers

Nonmetallic box

Hot

Neutral

Neutral

Travelers

Travelers

COM terminal

Four-Way Switching

A three-way switch must always be on each end of a multiple-point switching circuit. Taping the traveler white wire black to indicate it's hot is required.

Three-Way Switch at Entry

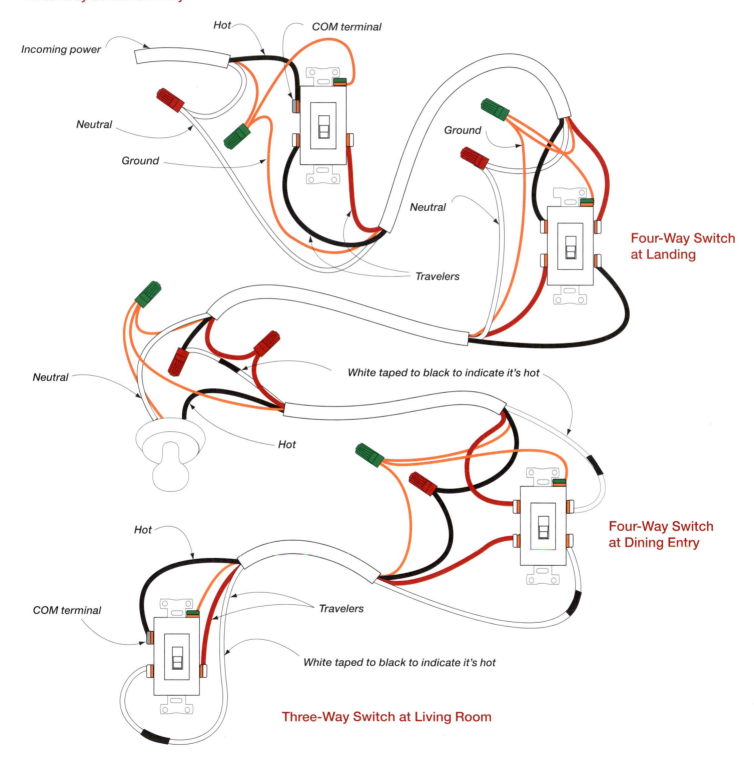

Hot

COM terminal

Incoming power

Neutral

Ground

Ground

Neutral

Travelers

Four-Way Switch at Landing

Neutral

White taped to black to indicate it's hot

Hot

Four-Way Switch at Dining Entry

Hot

COM terminal

Travelers

White taped to black to indicate it's hot

Three-Way Switch at Living Room

or a television at preset times so that people will think someone is home, even when the house is empty.

No matter what the use, however, most specialty switches are wired the same way and have the same two terminals that an ordinary switch has. The one thing to look out for is whether the switch is polarized, which means each wire must go to a specific screw or location on the switch. With a common snap switch, it doesn't matter which wire goes where.

Light Dimmers

Dimmer switches were developed to allow more control over lighting than just *on* and *off*. As a side benefit, they also save energy and extend bulb life by reducing the amount of electricity used. Dimmer switches are available in single-pole and three-way models. High-tech electronic multilocation switches allow you to dim the lights at a number of locations from one central dimmer switch.

Single-pole dimmers can be used for only incandescent lights. Never use a dimmer to control a ceiling fan or any other motorized appliance—you'll burn up the dimmer and the appliance. Although a table lamp will work fine, you will damage all other appliances plugged into that receptacle.

Wiring dimmers

Single-pole dimmers can be installed by simply replacing a single-pole switch. However, three-way dimmers cannot simply replace every three-way switch. Only one of the two three-way switches can be replaced with a dimmer; the other must remain a standard three-way switch.

Specialty switches come in many varieties. At left is a pilot light switch light (for a basement or attic light) and, at right, a timer switch (for a bathroom heater or fan).

For standard dimmers, the rough-in wiring is done the same way as for standard switches. If there will be only one switch, the two wires that would have gone to the standard switch simply connect to the dimmer. If a three-way system will be dimmed, choose the switch side on which to install the dimmer. If a four-way-or-larger system will be installed, the logic is the same—only one dimmer in the system. Dimmers with two black wires have no polarity, so it doesn't make any difference which wire gets hot or neutral. If the dimmer has a black and a red wire, the black normally goes to the incoming power source; the red, to the load.

Fluorescent dimmers

You cannot dim standard fluorescent lights because their ballasts are not designed for dimming. Instead, special electronic ballasts or magnetic dimming ballasts are required. Magnetic dimming ballasts work with T-12 fluorescent bulbs. This type of ballast uses large magnetic

■ **WORK SAFE**
■ **WORK SMART**
■ **THINKING AHEAD**

Never wire a dimmer hot. That is, never hook up a dimmer to an electrically hot wire—turn off all power first. The surge will burn up the dimmer's electronics.

Dimmers are quite easy to wire. Take the two wires that went to the switch to the two dimmer wires. Newer dimmers may have a ground wire. Normally, no polarity is observed.

components that may emit a buzzing noise, and the lamps may visibly flicker at low light levels.

Common problems with dimmers

No product is problem-free, and dimmers are no exception. Specific problems include noise, buzzing, overheating, and a short life due to surges coming from light bulbs that blow.

NOISE Dimmers are noisy—sometimes it's noticeable, sometimes it's not. They put radio-frequency interference (RFI) on the power line on both sides of the dimmer circuitry—the load and the entire branch circuit—through conduction and radiation. Any load on the dimmer branch circuit receives this noise. For example, you may hear it through

stereo equipment, broadcasting equipment, intercom systems, public-address systems, and cordless telephones.

There are several ways to eliminate or reduce this noise. The first option is to buy a dimmer with a built-in noise filter. Next, isolate the lighting branch circuits from the receptacle branch circuits. It's also important to make sure all ground connections are firm. A poorly grounded or ungrounded light on a dimmer circuit generates noise like a radio antenna. Lastly, look for dimmers that are advertised as being quiet.

LAMP BUZZ As the current to the lamp is being adjusted, the lamp's filament will buzz (it's more noticeable with bulbs above 100 watts). All lamps buzz to some extent, but the problem can be minimized by installing high-quality light bulbs. The difference is in the filaments—for instance, rough service bulbs have heavier filaments that reduce or eliminate vibration.

HEAT AND OVERLOADING This problem is typically created by designers or homeowners who create lighting systems that pull too much current through the dimmer. The solution is simple: Don't exceed the dimmer's wattage rating. This problem is extremely common with track lighting and with several light banks ganged together. If extra lights are added to the track lighting or if the bank's bulbs are replaced with higher-wattage bulbs, the dimmer can easily become overloaded.

A standard low-wattage dimmer uses its aluminum mounting plate as a heat sink. Because heat is the bane of dimmers, the thicker the heat sink, the better. Higher-wattage units, such as 500, 1,000 and 2,000, dissipate heat through heat sinks mounted at the wall plate. If dimmers are ganged together in the same

■ WORK SAFE
■ WORK SMART
■ THINKING AHEAD

When a bulb blows, a surge occurs on the line. This surge can take out a dimmer. Some manufacturers have addressed this problem by installing fuses in their bulbs, but there is no way to know which bulbs have them. It helps to buy better bulbs and a heavy-duty dimmer.

Although more expensive, the larger 1,500-watt dimmer (left) has a large heat sink, enabling it to handle more lights and last longer than the standard 600-watt dimmer (right).

box, some of the fins on the mounting plate may have to be broken off so that they can overlap, reducing the size of the heat sink. If this is the case, the wattage of each unit should be reduced by 20% to 50% (for example, a 1,000-watt dimmer should be reduced to 800 watts, or eight 100-watt lights).

Fan Speed Controls

There are two types of fan speed controls: fully variable, also called full range, and step control. Both types of controls may be used on a paddle fan, regardless of the number of speeds the fan has—simply set the fan's speed on the highest setting as it's installed. The wall-mounted speed controller can then vary the fan through all its speeds. Be sure to read the directions that come with the fan. A manufacturer may indicate a specific control that should be used with its fans.

Noise is a problem with fans as well. I remember the first variable-speed control I installed and how loud the humming noise was when the fan was on. I've been trying to find quiet controls ever since. Step controls normally do not hum, but they can be used with only one fan and have a maximum operating current of 1.5 amps. Fully variable controls are noisier but are able to control the full-range speed of a single fan or multiple fans wired in parallel. The maximum current depends on the model, but 2- to 12-amp models are pretty common. As a fan's speed is reduced, the fan itself can emit an annoying hum. Make sure the controller addresses that issue as well.

CHAPTER 11

Wiring Fixtures

L IGHT FIXTURES help make a house a home, and each fixture has advantages and disadvantages that you should know about before installation. Again, planning is important. Even though fixtures are installed after the finished walls are up, you should decide beforehand which fixtures you need so that you use the correct rough-in ceiling box. Order the lights ahead of time and keep them on site so that you can check the base depth and fixture height to verify that they will physically fit in their locations.

Choose the Right Box

In general, ceiling boxes are round, not rectangular like receptacle boxes. Old-style boxes are 3½ in. in diameter, while the new-style boxes are 4 in. in diameter. The basic difference between a round box and a rectangular box is the size of the screw that holds the light fixture to the box. Round boxes use a thicker screw (a #8 rather than a #6 for common rectangular receptacles or switch boxes) and thus can support a heavier weight (fixture) before ripping out the box's female threads or stripping the threads on the

screw. However, for lightweight fixtures that weigh no more than 6 lbs., such as smoke alarms and small outside light fixtures, rectangular boxes may be used.

ABOVE CODE Now that you know to use round ceiling boxes to support heavy lights, the next logical question is how much weight can they hold? The answer: 50 lb. Attach lights heavier than 50 lb. directly to the joist and use the outlet box only to house the wire splice. The only exception is when the manufacturer specifies that the box can support more than 50 lb. However, it's been my policy not to hang a light on an outlet box if it weighs more than 25 lb. If the screws pull out of the box and the irreplaceable light fixture is destroyed—or falls and hits someone on the head—you don't want to find yourself arguing about why it wasn't your fault.

Many designs, many problems

There are many types of round ceiling boxes from which to choose (see the drawing on the facing page). Standard cut-in boxes are made for renovation work but are limited to lightweight fixtures. Thus, if you want to hang some-

Ceiling Boxes

Bracket is nailed onto the joist.

Bracket is nailed onto the joist.

Integral-nail boxes are inexpensive but break easily.

Lip leans into the edge of the joist for additional support.

Bar-Hanger Box

Box should not be used with a heavy fixture because it will sag.

Bar allows the box to be attached anywhere between joists.

Cut-In Boxes

Adjustment screw tightens the metal ears against the ceiling.

Finished ceiling is sandwiched between metal ears on the back of the box and the drywall ears on front.

Adjustment screws tighten the swivel ears against the ceiling.

Box is screwed directly to a ceiling joist or other wood member.

Pancake Box

Three Ways to Use Bridging to Support a Ceiling Outlet Box

*If you can't attach the box to a joist, install 2×6 bridging
or a fan-mounting bracket between joists.*

Joist

2×6 bridging

Joist

Finished ceiling

Standard box with
metal brackets

Joist

Joist

Finished
ceiling

Outlet box removed
from bar-hanger box
and attached to bridging.

Joist

Joist

Adjustable
fan-mounting
bracket used
for light fixture

Finished ceiling

thing heavy, you'll have to cut into the ceiling and attach a support. Bar hanger boxes are commonly used in renovation and new construction. These use 16-in. to 24-in. bars to span joists, allowing the box to slide anywhere along the bar. However, they tend to sag with even moderate weight, and I don't use them to hang anything. The more common boxes use captive nails or brackets for a fast, easy installation. But those have problems, too: one too many hits on a captive nail, and the plastic arm snaps. And the nails bend easily. If you are nailing into oak or locust, forget it— you'd better predrill before nailing. That leaves boxes with brackets. Such boxes come in both narrow and wide support brackets. Always use the wide ones, which are much more stable and provide a more secure attachment. Typical ceiling boxes, regardless of design, have volumes ranging from around 13 cu. in. to 23 cu. in. Always try to buy the deepest ones.

Instead of forcing an overhead light to be only where a joist is located, you can install the light fixture wherever you want. If you are using a bracket box, nail a crosspiece on edge from joist to joist and attach the box to it. Even better, nail a scrap 2×6 or 2×8 flat across the joist span, remove the box from the bar-hanger, and screw it directly into the wood (use two screws). Before nailing the board into place, remember to recess it into the joists at a depth that allows the outlet box to be flush with the finished ceiling. Lastly, an adjustable fan-mounting bracket can be used to hold a light fixture (as well as a fan) between joists.

Metal pancake boxes are used to support light fixtures when you simply don't have the physical space to install a deep box, such as when the box must be surface mounted. Pancake boxes are normally ½ in. thick and 3¼ in. to 4 in. in diameter, with volumes ranging from 4 cu. in. to 6 cu. in. Light fixtures attach directly to them. When using this box, make sure that you do not create a cable-fill violation, which is easy to do because of the box's shallow thickness and low volume—it can only take a single 14-gauge cable. If volume is a problem, use a box with knockouts for the cable, not a box with cable clamps built in, to gain a bit of volume.

Pancake boxes are also used to support light fixtures with domed canopies, which are usually seen on the outside of a house adjacent to entries and exits. They're also commonly used to mount on solid wooden supports, such as exposed beams. Drill straight through the beam so the cable enters the center of the pancake box. The problem here is that sometimes the light fixture's base is not deep enough to cover the box. If this is the case, you will have to rout out or chisel a hole for the box to sit in.

Wiring an Incandescent Light Fixture

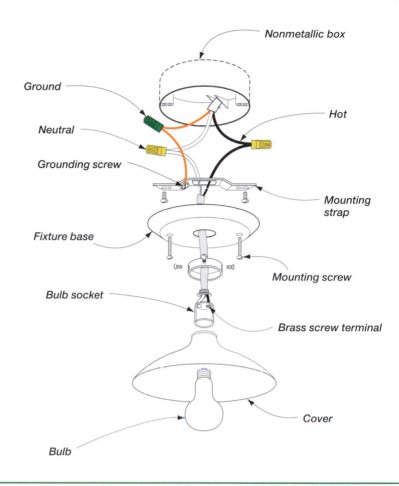

Nonmetallic box

Ground

Neutral

Grounding screw

Hot

Mounting strap

Fixture base

Mounting screw

Bulb socket

Brass screw terminal

Cover

Bulb

Incandescent Lights

I do not care for incandescent light fixtures that totally encase the bulb. The heat generated by the bulb transfers to the light fixture and sometimes fries the wiring—especially if you install high-wattage light bulbs. Most standard incandescent overhead light units are 60 watts maximum per bulb. If you need more illumination, look for a light fixture that is UL rated for higher wattage. In addition, look for a light that is easily disassembled for bulb replacement. Standard bulb bases are sturdier than small, fancy, specialized bases. Avoid large, round bulbs with extremely small, threaded bases because they snap easily at the base/bulb junc-

tion. When installing the fixture, do not throw away the insulation above the base (as is commonly done because it gets in the way of the attachment screws), because it keeps the heat of the fixture from coming through and frying the wires and wire connectors.

Rough-in installation consists of mounting the box and inserting the cable to power the light. After the walls are painted and the fixture is added, attach the incoming neutral conductor to the silver screw or white wire, and then attach the incoming hot conductor to the brass screw. A ground wire or screw on the fixture connects the metal frame to the equipment-grounding conductor.

■ WORK SAFE
■ WORK SMART
■ THINKING AHEAD

Stay away from any fixture that uses plastic female threads to screw the bulb into the fixture. To make connection to the bulb, they have a tiny metal strip in one of the female threads. These tend to make the bulbs flicker or not burn at all because of poor contact with the metal strip.

Fluorescent fixtures use different bulb designs depending on the bulb's length. These include (from the left): high output (this is what I have above my computer desk), dual pin (used in 4-ft. and smaller bulbs—I hate these), and single pin (very dependable).

Fluorescent Lights

Fluorescent lights provide a lot of illumination at a low cost and are therefore very popular. Standard fluorescents normally provide sufficient illumination, but if you require more, you can buy high-output and very-high-output fluorescents. Some typical problems with fluorescent fixtures include flickering and buzzing. In both cases, the only solutions are to change the ballast, which sometimes doesn't work, or install an expensive electronic ballast. Standard ballasts sometimes have a short life span and may short cycle when they become hot, cutting out the lamp until the ballast cools down. Standard ballasts are not meant for cold weather, either. If you plan to install fluorescent fixtures in an unheated shop, garage, or basement, you may have to buy cold-weather ballasts. Corrosion is a problem in damp areas, such as a garage or basement. Both the male pins on the bulb and the socket will corrode over time and produce intermittent problems. Lastly, some units seem to be so cheap that they have loose connections from day one.

Fluorescent lights come in either enclosed or bare-bulb fixtures. Your choice will be based on aesthetics: Enclosed fixtures are more attractive than bare-bulb fixtures, but the plastic enclosures may yellow over time. The most common fixture lengths are 2 ft. and 4 ft., and these fixtures take bulbs with two pins on each end. For really large rooms, you might consider installing 8-ft. fixtures. The bulbs for 8-ft. fixtures have a single pin on each end, which makes them much easier to change than the smaller two-pin bulbs. Personally, I prefer 8-ft. units no matter what the room size, because there are too many problems associated with 4-ft. fixtures.

Installing fluorescent fixtures is easy. For a hard-wired unit, simply take the NM cable through a knockout in the box (use an NM connector) and screw the fixture into the ceiling joists or a flat piece of wood installed between the joists (drill a large hole through the wood to run the cable). Wiring the fixture is simple: black to black, white to white, and ground to frame. Some fixtures are equipped with a plug and cord. These units require a standard 120-volt receptacle.

Never attach a fluorescent fixture directly to the grid frame of a drop ceiling. If the ceiling frame bends even a little, the fixture will crash to the floor (I've seen it happen). Fluorescent fixtures are heavy and should be independently and securely mounted to ceiling joists or 2×6 bridging.

Under-cabinet kitchen lighting

Kitchen countertop lighting often features fluorescent lights mounted under the cabinets. Such illumination removes the shadows that are invariably present when using ceiling lights. Try to get a fixture that is narrow—about 1½ in.—so that it can be hidden. Such fixtures normally have built-in switches. Some can be hard-wired and others come with a plug and cord. A plug-and-cord fixture requires a 120-volt receptacle nearby. For a hard-wired unit, pull the rough-in wiring to just under the kitchen cabinets and mount the lights against the wall so the wiring isn't exposed. Hookup is the typical black to black, white to white, and ground to frame or green wire. The units also can be butted up against each other for a continuous lighting arrangement.

Both open-bulb and covered-bulb designs are available. Open bulbs provide more light, but if the bulb breaks, the glass can go into your food. At my house, I installed covered lights but wound up taking the covers off to get more light—I decided the bulb-breaking situation was too remote to worry about. Low-voltage halogen lights are also available for under-cabinet lighting, but they are better suited for task lighting.

Buying Fluorescent Lamps

If you're tired of the harsh, gray light given off by standard fluorescent bulbs, look for a lamp with a higher Color Rendering Index (CRI). The CRI rates light sources on a relative scale from 0 to 100 (sunlight = 100). Lamps with a higher CRI make people and objects look more realistic.

A standard 34W "cool white" lamp has a CRI score of 62, which, to me, looks like a depressing yellow. Lighting stores and home centers also stock "daylight" lamps; some of them are rated up to 90. The only downside is price: Daylight bulbs cost two to three times more than standard bulbs.

Ceiling Fans

Ceiling fans have been very popular since Hunter started making them in the late 1800s. Hunter took advantage of the fact that blowing air down into a room makes people feel cooler in the summer. (Moving hot air also helps keep away mosquitoes—another big plus.) In addi-

Wiring a Fluorescent Fixture

Note: The integral wiring has been omitted for clarity. Neutral may wire to ballast depending on design.

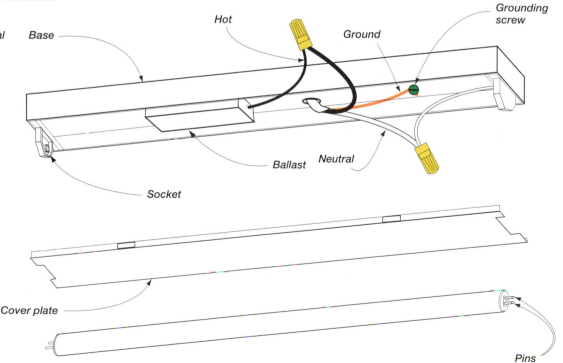

The bulb mounts outside of the cover plate for easy replacement.

Base, Hot, Ground, Grounding screw, Socket, Ballast, Neutral, Cover plate, Pins

Summer Cooling, Winter Warming

Summer Cooling

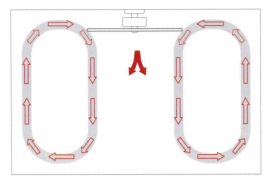

The fan cools by pulling air up along the exterior walls, bringing it to the center of the room, and then forcing it down.

Winter Warming

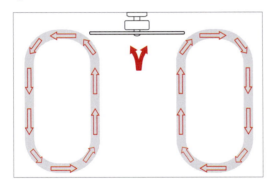

With the fan in reverse mode and set on low speed, warm air trapped at the ceiling is circulated downward along the exterior walls.

tion, reversing the motor to pull air up to the ceiling circulates warm air down exterior walls, making people feel warmer in the winter.

Before you think about installing a fan, you need to consider a few things. First, you must make sure that you have enough overhead room. Don't put a ceiling fan in a bathroom, where people will raise their arms to undress or dry off. You need at least 7 ft. of clearance from floor to blades; any less and you'll be giving people crewcuts as they walk by. For a low-ceiling room, you can buy a special kit that allows the fan to be installed flush with the ceiling. And you'll also need at least 18 in. from any wall to the ends of the blades for efficient air movement. Another important consideration is the ceiling pitch. The fan must hang down far enough so that the blades don't hit the ceiling.

Fans range from about 32 in. to around 52 in. in diameter. To get the most air movement and comfort, buy the appropriate-size fan for the room in which it will be installed. Here are some guidelines for rooms with 8-ft. ceilings:

- For rooms up to 64 sq. ft., buy a 32-in. fan.
- For rooms up to 144 sq. ft., buy a 42-in. fan.
- For rooms up to 225 sq. ft., buy a 44-in. fan.
- For rooms up to 400 sq. ft., buy a 52-in. fan.
- Larger rooms may require more than one fan.

I don't recommend buying a ceiling fan with light fixtures unless you first know the problem you may have with fan wobble. Most fans wobble a little; adding a light fixture accentuates the problem; however, many times the wobble doesn't develop until a year or two later. Why does this happen? Fans from the factory are almost perfectly balanced, but if you put a light kit on it and the light is imbalanced—or you accidentally bend one of the blades—the fan may wobble.

The bottom line is that a good quality fan/light combination should not wobble, provided that it is installed correctly. Make sure that the fan outlet box is securely attached via screws to a secure mounting, such a joist, and that the fan is tightly mounted to the outlet box. Lastly, be sure to assemble the light per the instructions, tighten all screws, and do not bend the blades.

Buying guidelines

Try to buy a fan that's designed intelligently. Make sure the fan allows you to hang it at an angle while you wire it, and then, once it's wired, swing it up into place. Nothing is worse than trying to wire a fan that has no swing-up bracket. In the old days, I hung a fan from a homemade rig while I wired it and then removed the rig to install the fan once I had the splices made.

Good fans need not be the most expensive on the showroom floor—just don't get the cheapest ones. For example, better-quality fans have sealed bearings so dust can't get in—cheaper ones don't (some are shielded on one side only). In my house, I installed three $50 paddle fans and the bearings wore out on all of them within 5 years (I live on a gravel road and dust is a problem). Here are some other things to consider:

- Low-cost fans have cheap, low-wattage motors (some of which are advertised as energy conserving). To compensate for this, the pitch of the blades is reduced (made more parallel to the floor) so that the motor doesn't burn out. As a result, not much air is moved. Look for pitches between 13 and 25 degrees. Blade length and the motor's rpm also affect air movement. All this totals up to cubic feet per minute, or cfm, of air moved. Look for the highest cfm listed on the box.

- Check the box for bragging points, such as "double the size of most fan motors," "sealed bearings," "sturdy steel motor," and "quiet operation." Also, listen to the display models. If it sounds like the motor is humming, keep looking.

- Look for units that have sealed, permanently lubricated oil-less bearings. Cheaper exposed bearings collect dust and wear out quickly.

- Look for a fan from a high-quality, brand-name manufacturer that comes

Existing ceiling fans can be wired to a remote control. This unit attaches under the mounting hood. Leave the antenna wire (third from the right) visible; otherwise, the unit may not work well.

with a long warranty—so that you never have to use it. The problem with most warranties is that they're so restrictive. Not having the original receipt and/or the box it came in can sometimes void the agreement. And don't forget the warranty doesn't cover labor. Still, the warranty provides some indication of the manufacturer's faith in its product.

Fan design can add to the theme of a room. Don't be satisfied with a standard design if you want something different. I changed one of my typical ceiling fans to a replica of an 1886, cast-iron, oval design. Manufacturers make a wide variety of fan styles, including replicas of antiques and even cartoon themes. They also make fans for wet environments, such as an outside porch.

Hanging a fan

When installing a ceiling fan, follow the manufacturer's instructions. For maximum comfort, position the fan as close as possible to the center of the room. The fan must be mounted securely via an approved plastic or metal box (I prefer metal because it is more stable). The inspector may look for an approval sticker as well.

ABOVE CODE   I install metal UL-approved boxes for paddle fans for all overhead light fixtures in bedrooms and living rooms—whether or not the homeowner is installing a fan. That

Supporting a Heavy Fan

Lag screws through the box into the ceiling joist.

U-hook

Hot to fan

Pancake box doesn't support the fan, so it does not have to be listed for such purpose.

Pin

Neutral

Grommet

Hanger bracket

Ground

Hot to light

A fan weighing more than 35 lb. must be supported independently of the ceiling box unless the box is specifically approved by the manufacturer.

Wires to the fixture pass through the drop pipe.

Canopy is omitted for clarity.

Two Ways to Wire a Fan/Light Combination

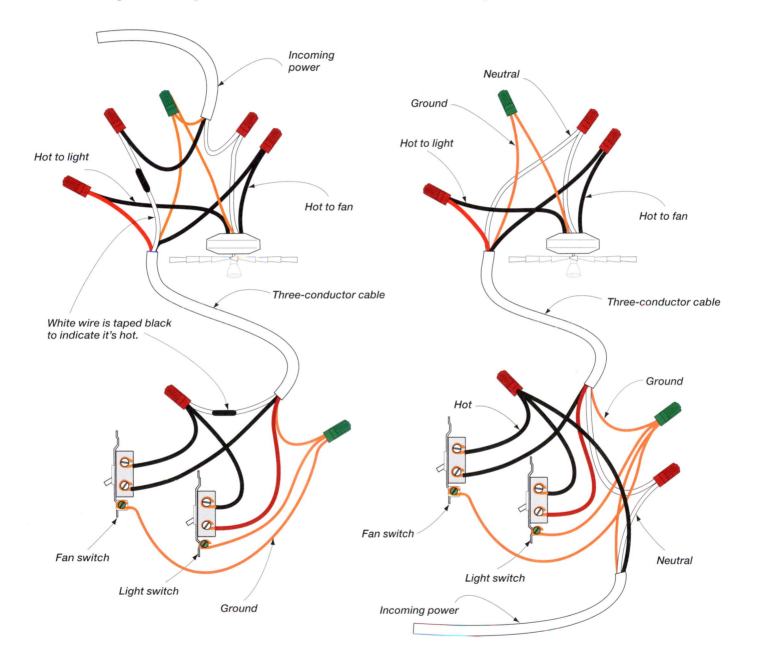

Not Preferred

Power is brought to the ceiling box, then to the switch.

Incoming power

Hot to light

Hot to fan

White wire is taped black to indicate it's hot.

Three-conductor cable

Fan switch

Light switch

Ground

Preferred

Power is brought to the switch, then to the fan.

Neutral

Ground

Hot to light

Hot to fan

Three-conductor cable

Hot

Ground

Fan switch

Light switch

Neutral

Incoming power

way, if future homeowners decide to install a paddle fan, they'll already have the right kind of box in place. Fan-approved boxes are also easier to install during rough-in than as a retrofit.

No matter which box you use, make sure it is securely attached to a stable ceiling joist or cross support. If the joist wobbles, so will the fan. If a joist is not close to where you need to install the fan, nail a 2×6 bridge from ceiling joist to ceiling joist and mount the fan to it. Approved fan-mount kits are also available so that you can mount a fan from joist to joist without having to go into the attic.

The weight of a ceiling fan is an important consideration. First, the box must be designed and listed especially for the purpose of fan support. Second, only fans that weigh 35 lb. or less, with or without accessories, can be supported directly from a ceiling box. In most cases, fans that weigh more than 35 lb. must be supported independently of a ceiling box unless the box is listed by an agency for increased weight. (Some boxes can support up to 70 lb.) For such an installation, the fan manufacturer supplies a U-hook that screws into the wood beam in the ceiling. A rubber grommet fits through the center of the U-hook and a hanger bracket, attached to the down rod, fits over the grommet pins on the outer edges of the grommet (see the drawing on p. 204).

Wiring a fan

Wiring a ceiling fan is quite simple. Just follow the manufacturer's instructions. There is a white neutral, a green or bare ground, and two colored wires—one for the light (if there is one) and the other for the fan (normally black). If you want to control both independently, the rough-in wire must have three conductors plus a ground—a three-way switch cable. The power cable can be brought to the fan, but I prefer to take it to the switch box for easier wiring and troubleshooting. Also, do not use a light control or dimmer as a variable-speed control for the fan motor. Instead, use a speed control that's designed for ceiling fans.

Some fans come with a remote control; others you wire the remote later. Either way, a remote is a nice thing to have, because you can control lights and fan speeds without getting out of bed.

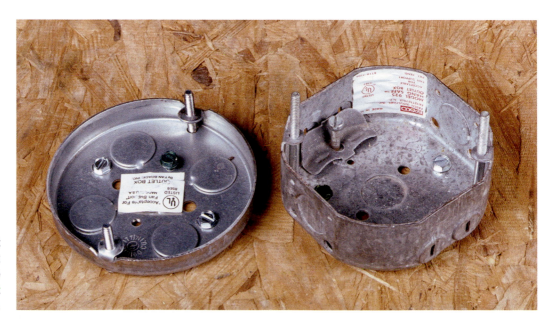

Fan-mounting boxes must be UL approved. The box on the left is a pancake box. The one on the right is a standard ceiling box.

Smoke Detectors

When installed properly, smoke detectors save lives. Smoke detectors are required by code to be installed in the hall outside bedrooms. This does not save you if the fire is in the bedroom. Ideally, smoke detectors need to be both inside and outside bedroom doors (this is sometimes required by building codes). If installed in the bedroom, the circuit must have AFCI protection (see pp. 159–161).

Detectors must be hard-wired for new construction, and many municipalities also require a battery backup system within the unit. Most areas allow battery units for houses already built when the homeowner just wants to have some protection. Code also requires that all hard-wired units alarm at the same time. To do this, all units must be wired together with three-conductor wiring—a three-way switch cable, normally 14-3.

Wiring a Smoke Detector

Outgoing power (14-3 cable)

Joist

Ceiling box

Incoming power (14-2 cable)

Ground

Neutral

Red wire triggers alarms on other units downstream.

Hot

Mounting strap

White and black wires provide power to this unit and to others downstream.

Test button

Track Lighting

Track lighting can provide accent lighting in various areas with only one cable bringing power. The problem with track lighting is that the lamps for one track may have a number of different wattages, which can quickly add up. It's very easy to go from a simple 60-watt circuit to over 1,000 watts by simply plugging a few more lights into the track. Thus, it's easy to overload a circuit. There is no set per-foot wattage in residential housing; for commercial wiring, however, it's 180 watts per 5-ft. section (nonsimultaneous use) and 180 watts per ft. (simultaneous use).

Track lighting needs a flat mounting surface. This is not a significant problem with one short piece of track, but it is a problem with a long track or several short tracks if the ceiling has both low and

high spots or a sprayed-on compound to make it look wavy. In these cases, you may see gaps between the track and the ceiling or wall.

Rough-in wiring for track lighting needs to be as precise as possible to make sure the cable enters the exact spot where the wires insert on the narrow track. Sometimes the rough-in wire can terminate in a standard outlet box. Most times, the splice is done in the end of the track or in the connector, so the cable must be brought through the finished wall at the exact spot where the track begins or ends. Make a hole no larger than the diameter of the cable through the finished wall. The power wires connect to screw terminals within the track.

There is not much room inside the narrow track for a lot of wires but the wiring logic is still the same: black to the brass screw, white to the silver screw, and bare wire to the frame.

Outdoor Light Fixtures

The most common outdoor light fixtures on a home are flood lights, entry lights, dusk-to-dawn lights, motion detectors, and low-voltage landscape and security lighting. The wiring for these fixtures is pretty basic but installation can be complicated. In some cases, planning the locations can save a world of trouble later.

Floodlights

Floodlights come in wattages ranging from 75 to 1,500 and are normally mounted outside at a high point on the house. Standard flood-lamp bulbs can be mounted up to three in one place, all facing different directions for maximum coverage. Don't mount these lights in an inaccessible location because you'll need to change the bulbs at some point. If possible, mount the light above or adjacent to an upstairs window so that you can easily change the bulbs. Always point the bulbs down to keep water from running into the base. Try to find bases with copper threads, rather than aluminum ones, to keep the bulb threads from seizing to the base threads. If you can't find bases with copper threads, lightly wipe an anticorrosion compound onto the bulb's threads. Quartz lights can also be used for outside lighting, but I don't recommend

them. These fixtures get hot enough to burn the skin right off your fingers and the bulb's life can sometimes be measured in minutes (another one of my mistakes that will save you time and money).

For rough-in, simply bring the cable out of the building. Install a watertight box and the fixture after the finished exterior is up. If you try to install the box ahead of time, you may not be able to match the exterior surface, and you will spend a lot of time trying to place a cover on the box that doesn't conform to the finished wall.

Entry lights

Entry lights can be troublesome if the building has beveled or irregular siding. If the fixture is placed flat against the siding, the light will slant along with the siding. If placed plumb to the wall, the light base will have a large gap underneath it because of the bevel of the siding. In this case, it's best to have the carpenter make a fancy base and install it for you. If that's not possible, grab a piece of siding and invert it on the installed siding to make a flat area. If the house has vinyl siding, buy an adapter that allows the fixture to seat on the exterior with a minimum of problems.

Once you've got a flat mounting surface, install a pancake box to mount the light. Double-check the light base to see if it is deep enough to hide the box—if not, recess the box. Roughing in simply involves sticking a cable through the exterior wall where the light will be mounted.

Dusk-to-dawn lights

Normally installed as security lights, dusk-to-dawn lights can be bought as an entire unit or as individual lights. Dusk-to-dawn lights normally have a plug-in photoelectric cell, or eye, which senses when it's dark and then turns on the light. These units are normally mounted in areas for parking or security. Packaged units normally use mercury vapor lights, while individual sensors can turn on a light for any purpose, such as a sign to be lit after dark. These cells don't last forever (neither do the bulbs) so locate the fixtures where you can easily reach them when it's time to replace the bulb or cell.

Typical wattages range from 70 to 175. Lower wattages may be cheaper, but they may not throw enough light. These lights are normally hard-wired on a 120-volt circuit. The disadvantage of such a fixture is that it could take several minutes for the unit to turn on. Separate photoelectric cells can be wired in series with any light to turn on after dark, but make sure the cell matches the wattage of the bulb, or it could burn out.

Motion detectors

Motion-sensing devices are very popular nowadays. Some units detect just motion, while more sophisticated models use infrared rays to detect body heat. The units turn on house lights or flood lights as someone comes near the house. You can also adjust the sensitivity of the unit to compensate for dogs, raccoons, and other small animals. Loads for such devices are typically limited to around 500 watts. You can buy them as individual sensors, which turn on lights that are already installed, or as complete sets with a light-attachment assembly.

Installing such a device can be problematic. First, make sure the unit you buy can detect at the distance you want—30 ft. to 60 ft. is common (most units have trouble detecting motion at farther distances). Mount the detector no higher than 12 ft. above the ground. The higher the unit is installed, the less ground surface will be covered and the less responsive the unit will be to motion on the ground. Position the unit away from heat-producing sources, such as a heat pump, and away from reflective surfaces, such as windows and pools. Trees and ferns that wave with the wind will also create problems.

It may seem logical to face the unit in the direction of the object that will turn the unit on, but this is not what you want to do. These sensors—both infrared and motion—do not saturate the area. Instead, they project fingerlike beams (called lobes). If you're facing the unit, you can sometimes slowly walk toward it (between the detection lobes) and not set it off. The directions on the unit tell you to install the unit at right angles to the area you want to track so that people will pass across the lobes of detection, not between them (see the top drawing on p. 210).

The rough-in wiring only needs to be a cable sticking through the wall. Attach a watertight box to the wall after the siding is up and pull the cable through. Make sure the box shape matches the detector's shape (you don't want a round box for a square detector fixture). Once the box is up, simply follow the wiring instructions included with the unit.

Low-voltage landscape and security lighting

Low-voltage landscape and security lighting has become popular in the last few years. Low voltage means it runs off a

■ **WORK SAFE**
■ **WORK SMART**
■ **THINKING AHEAD**

Avoid motion detector units that don't come with instructions (there is at least one) or units that don't have sensitivity adjustments. Also, avoid units with a quartz light. The glass front becomes too hot and the light tends to die an early death.

Locating a Motion Detector

Mount the motion detector no more than 12 ft. above the ground and locate it so that motion cuts across the lobes of detection. Locating the unit straight into the target area may allow someone to walk up to the house undetected through a dead zone.

Not Ideal
Detection lobes extend straight into the target area.

The walkway is the target area.

Detection lobes

Dead zones

110 degrees

Motion detector

Ideal
Detection lobes cut across the target area.

The walkway is the target area.

Detection lobes

Dead zones

110 degrees

Motion detector

Wiring a Typical Motion Detector

Hot

Ground

Watertight outlet box

Hot to sensor

Neutral

Gasket

Plastic

Sensor controls hot output to lights.

Do not exceed maximum recommended bulb wattage.

transformer (plugged into a wall outlet) that provides around 12 volts to power a string of lights. Compared to their 120-volt cousins, low-voltage systems are safer to use and easier to install. In some cases, installation requires little more than stabbing the lights into the ground. The cable can be laid just under some mulch or buried under a few inches of lawn. Such lighting is generally suited for short distances (less than 100 ft. from the transformer).

You cannot plug a transformer into an outside outlet and leave the outlet's door open. You must use a watertight-while-in-use outlet or outlet cover. If the area to be illuminated is a considerable distance from the house, bring the AC power to the location via a UF cable and install an outlet onto a post.

Although generally economical, landscape and security lighting can get expensive if you buy fancy lamps.

Commonly available path lights come in the shapes of mushrooms, tulips, shells, pagodas, mini bulbs, and micro bollards (shaped like a pencil with 360 degrees of illumination); they can also provide indirect light (reflected off the bottom of the fixture). These fixtures run from $50 to $100 per light. If you want landscape and security lighting, follow these procedures:

- Determine what will be illuminated (trees, walks, etc.).
- Determine the type of illumination (spot, shadow, silhouette, overall, etc.).
- Determine the number and location of lights (hidden in trees, spacing along walk, etc.).
- Determine the types of mounts (stakes, outlet box, building into structure, etc.).
- Determine bulb wattages (27 to 50 watts is typical).

- Add the bulb wattages together and add 10% to select the transformer. Go to the store to see if they have a prepackaged kit that meets your needs. Kits come with properly gauged cable. If you buy everything separately, opt for the largest-gauge cable so you can add more lights later.

Recessed Lights

Recessed lighting provides a variety of illuminating effects throughout the house, adding light to an area without affecting other design criteria. It is commonly used in kitchens and closets and for accent and task lighting.

The housing is the main factor in a recessed-light installation. You can buy housings for new construction or remod-

Insulation and Recessed Fixtures

T Housings
Keep insulation 3 in. from fixture.

ICT Housing
Okay to cover with insulation.

Wiring a Recessed Light

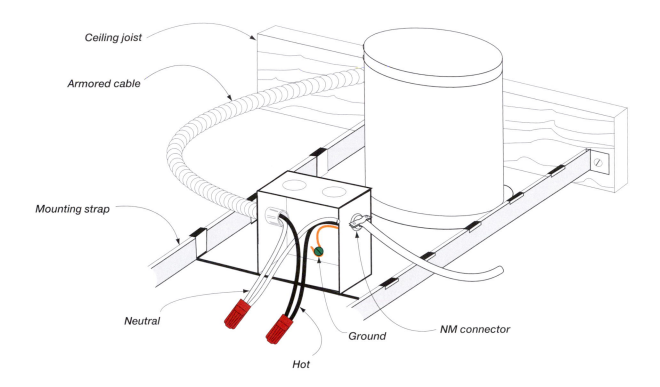

Ceiling joist

Armored cable

Mounting strap

Neutral

Hot

Ground

NM connector

■ **WORK SAFE**
■ **WORK SMART**
■ **THINKING AHEAD**

If a recessed light occasionally turns itself on and off, it's probably due to the heat sensor. This frequently happens when the wattage of the bulb is too high. Look at the sticker inside the can to learn which bulb designs and wattages can be used. If you haven't exceeded the wattage, try installing a lower-wattage bulb.

eling. There are two basic types from which to choose: T-housings for uninsulated ceilings and ICT-housings for insulated ceilings (commonly called non-IC and IC). Non-IC housings must be kept at least 3 in. from insulation in the ceiling or walls. IC housings are preferred, because you can insulate around them and they normally have a lower wattage. All housings are equipped with a thermal protector that shuts the light off if insulation material is installed too close to the housing or if the lamp wattage is exceeded. The type of housing is written on a sticker where the bulb screws into the socket.

IC fixtures have an ICT-housing (standard round), a UICT-housing (double-wall rectangular), or an ICAT-housing (airtight). Housings are also available for special applications, including sloped ceilings and closet lights. There are also housings with deep baffles designed for superior visual cutoff and others for high-wattage lamps (up to 200 watts). For remodeling jobs, make sure that there is enough room in the ceiling height for the fixture. If height is a problem, shallow models are available. Use only trim that is compatible with the particular fixtures you're installing.

Wiring is normally done via a metal splice box that is attached to the housing frame. Rough-in includes mounting the housing but not the trim plate. When mounting the housing, be sure to allow for the finished ceiling (in new construction).

CHAPTER 12

Wiring Appliances

OUR HOMES are becoming more appliance-oriented—and the trend doesn't seem to be slowing. With the increasing number of appliances in a house, wiring is taking more of an electrician's time on the job site.

Gone are the days when you simply plugged in a new appliance. Today, larger, wattage-hungry appliances might require their own circuits. Others must be hard-wired, meaning that the branch-circuit wire is connected directly to the appliance. Also, voltage varies: Some appliances are 120 volts, others are 240 volts, and still others are both 120 and 240. And they all wire differently.

Kitchen Appliances

The kitchen typically houses the greatest number of electric appliances, including a dishwasher, refrigerator, garbage disposal, microwave, stove, and a slew of others. Thus, it has the greatest amount of wiring and wiring problems. Here are some approaches to wiring appliances.

Dishwasher

Residential dishwashers are normally 120 volts and require a dedicated 20-amp branch circuit using 12-gauge wire. From behind the dishwasher, the cable is brought through the lower part of the wall or up from the floor adjacent to the wall, and then (under the dishwasher) to a metal splice box located in the front just behind the kick panel.

The first step is to install an NM connector, or strain relief, on the splice box,

Wiring a Dishwasher

Kick panel

and then bring in the cable. The dishwasher's wires come into the splice box via a prepunched hole. Wrap a few turns of electrical tape around the wires at that spot. The wire connections are simple: hot to hot (black to black), neutral to neutral (white to white), and ground to box or the appliance's green wire. If the house wiring is old and has no ground, you'll need to run an equipment-grounding wire to the main panel or install a new cable. Never install a grounded appliance on ungrounded wiring.

Garbage disposal

Garbage disposals are wired from the bottom and are either hard-wired or wired through a plug and cord. If the unit is hard-wired, the cable comes right out of the drywall under the kitchen sink or straight up from the crawl space through the bottom of the cabinet. Be sure to keep the wire out of the way of the plumbing. The voltage is usually 120 and most ⅓- and ½-hp garbage disposals do not need to have their own dedicated branch circuit. The electrical connection is made under the electrical cover, or plate, on the bottom of the disposal. The splices are made with wire nuts: black to black, white to white, and ground to the frame screw on the unit. Be sure to install a switch on the wall near the kitchen sink to control the power to the disposal.

Electric stove

Electric stoves are 120 *and* 240 volts. The burners and bake unit require 240 volts, but the timer, clock, and buzzer use 120 volts. The NEC provides specific

■ WORK SAFE
■ WORK SMART
■ **THINKING AHEAD**

Some residential dishwashers allow 14-gauge cable and 15-amp protection, but don't use them. Someday the dishwasher's replacement may be more heavy-duty, and, once installed, may overload the wiring and kick the breaker.

■ WORK SAFE
■ **WORK SMART**
■ THINKING AHEAD

When running the dishwasher cable, bring in enough slack so that the dishwasher can be pulled out and serviced. Make a large loop, and then circle it under the appliance to the splice box.

Wiring a Garbage Disposal

Bottom view of disposal with cover plate removed

Flexible metal conduit (Greenfield)

Grounding screw

Ground

Neutral   Hot

Incoming power

Single-pole switch

Hot

Neutral

Ground

To disposal

guidelines for wiring stoves, including choosing the right-size conductors.

For a wall-mount oven and a drop-in range or cook top, the NEC requires you to use the nameplate rating to determine the correct-size conductor. However, the NEC allows the typical slide-in stove to be derated. That is, if the nameplate rating is not more than 12 kilowatts, the rough-in wiring can be designed for 8 kilowatts. The NEC allows this derating because the organization feels it is unlikely that all elements will be on high, or full power, for an extended period of time.

ABOVE CODE I don't agree with the NEC in this instance. The organization has obviously not had a large Thanksgiving or holiday dinner when the stove ran full blast all day long. I've had to replace many 40-amp slide-in stove circuits with 50-amp circuits because excessive current kept kicking out the 40-amp breaker on those occasions. To keep this from happening to you and ruining your holiday festivities, design the stove circuit for 50 amps (use 6-gauge copper) instead of using the NEC's derating standards.

A standard slide-in stove plugs into a 50-amp female receptacle. The outlet box can be surface-mounted or recessed. I normally attach a 2× nailer between the studs, bring the cable through the nailer, and screw the receptacle to the nailer after the finished wall is up. A more complicated installation is to use a recessed box and install a flush-mount receptacle. Because of the large cable required, you'll have to use a 4 11/16-in.-by-2 1/8-in.-deep square box (42 cu. in.), put a large cover plate over it, and mount the receptacle to that (see the photo above).

Because a stove pulls both 120 and 240 volts, the neutral must be insulated

(this applies to hard-wired units as well). As of the 1996 code, three-prong receptacles, both surface and recessed, are no longer allowed. The ground and the neutral cannot be connected, meaning that the neutral must be an insulated conductor separate from the ground. Now you need a male plug and a female receptacle with four conductors: two insulated hots, one insulated neutral, and one ground.

If the electric range or oven is a drop-in type, there should be a pigtail of wires, called a whip, with flexible conduit hanging from the unit. These connect to the incoming power cable in a large splice box that you supply. Simply bring the four-conductor cable into the box to splice to the range's wires. If the whip is not already installed, do not be tempted to use a plug and cord on a drop-in cooktop (some models even come with labels warning against its use). Instead, make your own whip with THHN wires (12 or 10 gauge, depending on the load) and run it through a Greenfield conduit. You can use a leftover piece of wire.

RANGE HOOD AND FAN Range hoods can be almost any size, from 30 in. up to 72 in. wide. You can get everything from a standard economy hood (I'd avoid this type because it doesn't pull enough air)

Stove outlet wired via 6 AWG NMB copper. The cable contains an insulated white neutral, a bare ground, and two insulated hot conductors.

■ **WORK SAFE**
■ **WORK SMART**
■ **THINKING AHEAD**

If you are mounting the stove outlet to the floor, keep it dead center of the proposed stove location. This keeps the feet from hitting the outlet as it slides back and allows you to disconnect the plug simply by pulling the stove's bottom drawer.

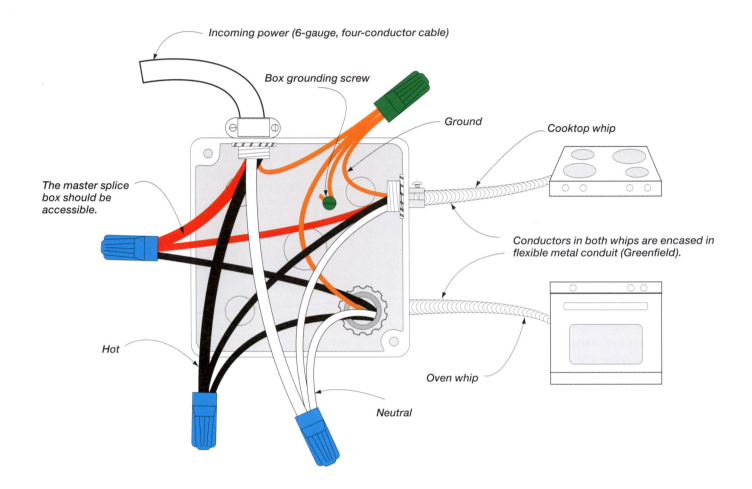

Incoming power (6-gauge, four-conductor cable)

Box grounding screw

Ground

Cooktop whip

The master splice box should be accessible.

Conductors in both whips are encased in flexible metal conduit (Greenfield).

Hot

Oven whip

Neutral

Hood Selection and Noise Control

The purpose of a range hood is to move air. The blower motors on economy models are noisy and move little air (around 140 cfm or less). Better quality units move 200 cfm to 400 cfm and are advertised by some manufacturers as being "quiet." Units with large dual blowers (up to 1,200 cfm) are available. If you don't want any noise at all (it is impossible to have the power you really need and keep the noise to a whisper), mount the blower on the outside of the house. Broan-Nutone, for example, has a roof- or wall-mounted 900-cfm blower for its Rangemaster® hood. This type of blower assist looks like an inverted aluminum bowl.

to a restaurant-grade canopy hood. Also available are high-tech models, with features such as blower speed control, heat-sensitive components that automatically turn the blower to high speed when excess heat is sensed, warming lamps to keep food hot, high-heat alarms, and dual lighting.

Most fans are 120 volts with the rough wiring connecting to a box within the unit. Because it pulls little current, the fan can be wired into the kitchen light circuit. The problem with wiring the stove fan is that, during the rough-in stage, the wire must be brought out of a bare stud wall before the fan and everything else in the kitchen is installed. Once the rough wiring has been run to the correct spot, staple it in place on the stud, leaving about 18 in. of cable sticking out beyond the stud wall. Then install the drywall and the kitchen cabinets. After that, mount the hood and fan unit to the bottom of the cabinets.

The rough-in wiring must be brought out exactly under the proposed bottom of the overhead kitchen cabinets, but not so far down as to be exposed once the fan unit goes in. If you run the cable too high, cut the drywall a bit to get the cable down to its correct location (you can do this only if the fan housing and overhead cabinets will cover the hole). If this isn't the case, you'll have some patching to do, which could get sloppy. The wiring is pretty simple: black to black, white to white, and ground to frame.

The standard installation height for fans is 54 in. to 60 in. from the floor to the bottom of the unit. Add another 5½ in. to 9 in. for the height of the unit. Try to bring the cable out of the wall about 7½ in. to the right of the duct's centerline. Make sure the drywallers do not move the cable location or cover it. If you're responsible for running the ductwork, make sure there is no stud dead center of the proposed location—you normally need at least 5 in. of clearance on each side of dead center.

Baseboard Heaters

Electric baseboard heaters are a fast and economical method of supplying supplemental heat. They are normally installed below windows and on adjacent walls. I can't tell you the number of units or the total wattage you'll need, because that depends on too many variables, such as

how cold it gets and how well the house is insulated. A general rule is to mount units under windows and have them run at least the length of the windows.

Electric baseboard heaters can be installed quickly and easily, as long as you know what to do and which problems could occur. The first decision to affect the wiring is whether the thermostats will be inside the heater or on the wall.

Both in-heater and wall thermostats are acceptable, according to the NEC, but I only recommend in-heater thermostats, with one thermostat per heater. An in-heater thermostat is more economical and faster to wire (wall-mounted thermostats literally double or triple the labor time of baseboard-heater installation). A wall thermostat turns all units on at once, which wastes a tremendous amount of energy. Heaters with integral thermostats turn themselves on and off independently to heat only the area needed, resulting in less overall power consumption. Having the thermostat along the floor also allows a faster response time to incoming cold air. I normally mount heaters 1 in. or so off the floor to allow room for the finished floor and to maintain an air gap between the two. Because they're really the only smart choice, I will limit this discussion to in-heater thermostats.

Although 99.9% of all baseboard heaters are rated to use 240 volts, 120-volt models are available. However, I don't recommend them, unless only 120 volts is available. Chances are high that the heater could accidentally be installed on a 240-volt circuit, resulting in a fire (this has happened). In addition, a 240-volt heater, which draws half the current of a 120-volt heater, is more efficient and requires less expensive wiring. I'll assume that all heaters in this discussion are 240 volts.

The wattage of the baseboard heater determines the heat you desire. Most

WORK SAFE
WORK SMART
THINKING AHEAD

Don't do what I did when I once attached a fan hood to a cabinet bottom. I used screws that were too long, and they went all the way through the cabinet bottom and into a fancy menu holder and some Tupperware. This is one of those errors you don't make twice—use short enough screws so that they don't poke through the wood.

WORK SAFE
WORK SMART
THINKING AHEAD

When selecting the location for a baseboard heater, remember that furniture cannot be placed in front of it, because the furniture will block the flow of air.

Smoking Heaters

When you first turn on a new electric heater, it may smoke a little—or a lot. Don't worry—it's only burning off the factory coating on the elements. This is normal and should stop after a few minutes. It's best to open the doors and windows and let the room air out when a heater is turned on for the first time. (Note that there will be more than enough smoke to trigger the smoke alarms.)

Older heaters also smoke if they have not been used for an extended period of time. Dust and cobwebs frequently collect on the elements. Remove the front cover of the baseboard (it should snap off) and vacuum the elements before turning on the unit.

Outlet Placement and Electric Heaters

NO!

Having a receptacle above the heater allows an appliance cord to fall in on the hot elements.

OK

Receptacle mounted in heater

OK

Receptacle mounted between heaters

panels are 250 watts per ft., or about 1 amp per ft. To figure the exact amperage, divide the wattage by the voltage. For example, say you're putting a 10-ft. unit (2,500 watts) under a picture window in a dining room. By dividing 2,500 by 240, you'll find that the amperage is just above 10.

Several small baseboard heaters can be fed from the same cable coming from the main panel as long as the current doesn't exceed 80% of the cable's rating: 12-gauge cable cannot take more than 16 amps of continuous current. This translates to a maximum of 16 ft. of baseboard per 20-amp breaker, 12-gauge circuit. Do not use 10 gauge (it is expensive and hard to run) or 14 gauge (it is too small).

In your desire to install the required number of receptacles, it's possible to have a receptacle located immediately above a proposed baseboard heater. This is not allowed. A cord that plugs into the outlet might fall into the hot burners and short out. Even if the cord doesn't fall directly into the elements, the heat will eventually damage the cord. To solve this problem, some baseboard-heater manufacturers have end, corner, and center sections that allow you to insert the outlet within the baseboard. However, the receptacle must not tap into the 240-volt circuit powering the baseboard. It must have its own 120-volt cable.

Wiring an in-heater thermostat

Installing the thermostat in a baseboard heater is quite simple. Simply pull off the cover plate on one end of the unit (it makes no difference which end) and mount the thermostat. Some mount to the heater, others to the removable panel. The rough-in wiring can be run directly to the heater's location and through the wall and the knockout in

Wiring In-Heater Thermostats

Line side going to thermostat

Hot

The white wire is taped black to indicate it's hot.

Load side coming off thermostat

Incoming power

Element

Ground

Single-Pole Thermostat

Low

Off High

Hot

Double-Pole Thermostat

The white wire is taped black to indicate it's hot.

Line side going to thermostat

Load side coming off thermostat

Incoming power

Element

Ground

Low

Off High

the heater (be sure to install an NM connector). You'll have a choice of a single- or double-pole thermostat (two thermostats ganged together). Go for the double. A double-pole thermostat cuts power to both legs, not just to one, and offers the convenience of an OFF position on the unit (single poles only turn down to low).

Once the cable is in the heater, splice one hot wire to each half of a double-pole thermostat or one hot wire for a single-pole thermostat; the other end of the thermostat connects to the heating element. This places the thermostat in series with one or both legs of the incoming power. If you splice the incoming power across the thermostat, a direct short will be created when the thermostat engages, and the thermostat will be ruined. (If you switch on the thermostat and hear a snap, crackle, and click of the breaker kicking off, and then see smoke, this is what you did...I won't tell you how I know this.)

Electric Wall Heaters

Electric wall heaters come with and without fans and in both 120- and 240-volt versions. Always buy the fan-operated unit because it better distributes heat around a room. This is especially important if there is plumbing in the room; because heat rises, you'll warm only the top of the room and allow the pipes at foot-level to freeze. Another rule: Always buy a 240-volt model, which draws half the amount of current that a 120-volt unit draws. You are also allowed to wire more than one heater in a branch circuit, as long as the wiring isn't overloaded. Do not exceed 16 amps on a 20-amp circuit. A typical 240-volt, 1000-watt heater pulls around 4 amps, while a 1500-watt unit pulls a little over 6 amps.

Always get a heater with an integral thermostat for the same reasons previously discussed (see Baseboard Heaters on p. 217) one cable, one hookup, and you're done. A small, single-wattage heater is ideal for a small room, such as a bathroom. Large, variable-wattage units with fans are better for heating large rooms or even an entire house, but heating a house with this type of heater only makes sense if the areas are open. The large unit is about 6 ft. tall and fits between studs or can be mounted on a wall. It can also heat an adjacent room if you install a length of duct from the heater to that room.

Typical in-wall bath heater right out of the box. The chrome front pulls out of the frame so that it can be mounted to the studs.

Utility-Room Appliances

Wiring the appliances in a utility room can be tricky. Many of those appliances are specifically covered by the NEC, and each situation is different. In general, the appliances in a utility room are large—water heater, heat pump, and dryer—and may use 240 volts only or both 120 and 240 volts. Wiring appliances this large is not a job for an amateur.

Water heater

Water heaters come in both 120 volts and 240 volts. Buy a 240-volt heater—it pulls half the amount of current and costs no more than the 120-volt models. For these reasons, I will focus my attention here on 240-volt heaters only.

Of all the appliances, a water heater is usually the simplest to install. The standard electric water heater is a 50-gal.,

Secure the frame on at least two sides. Extend the frame out from the stud the depth of the finished wall.

The power cable enters the frame's integral triangular splice box via an NM connector (bottom right). Tape the white wire black. The ground wire (not seen) attaches to a grounding screw mounted on the frame.

Once the finished wall is up, splice in the 240-volt heater. Attach one incoming wire to one heater wire—no polarity is required. Once spliced, stuff the wire connectors into the splice box and reattach the chrome front.

Wiring a Water Heater

Incoming power

Hot-water outlet

Cold-water inlet

Green grounding screw

Hot

Ground

Splice box

The white wire is taped black to indicate it's hot.

240-volt water heater

The white wire is taped black to indicate it's hot.

The cable from the cutoff switch to the water heater is encased in conduit (flexible in earthquake zones).

The cutoff switch can be a fused or nonfused disconnect rated for the load of the heater—normally 30 amps. It must be close to the heater.

240-volt, 4,500-watt, storage-type appliance fused at 30 amps. It is hard-wired with 10-2 w/g NM cable run from the main panel to the heater or to a disconnect adjacent to the heater. A disconnect is required if the heater is out of sight of the main panel, and the heater should be as close as possible to the switch. If the heater is in a habitable area, the NM cable from the switch to the heater must be enclosed in metal conduit (flexible conduit or Greenfield is best for this situation but not always required).

In older homes, the cutoff switch (not the heater) is frequently the problem when a homeowner complains that there's no hot water. If the water heater is in sight of the service panel containing its breaker, you can simply wire around the switch rather than replace it.

Wiring is quite simple, because there are only two connections plus ground and there's no polarity on the hot wires for 240-volt loads. Just bring the cable into the water heater via an NM connector and splice it into the two water heater wires via wire nuts. Tape the white wire black and attach the ground to the frame.

ABOVE CODE If you want to wire a 240-volt water heater with 12-gauge cable, the heater needs an element that is 3,500 watts or less. There is nothing wrong with using a smaller wattage element; it simply takes longer to heat the water. However, because the water heater has a label on its jacket that lists the specs of the unit, there is a possible safety risk. When the plumber or do-it-

Wiring a Dryer Receptacle

Violation

10-2 NM W/G

The white wire is hot.

10-gauge, three-conductor cable

Hot

Ground connected to the neutral terminal

Violation as of 1996 NEC

SEU

Hot

Hot

Stranded neutral

Meets Current NEC Guidelines

10-3 NM-B W/G

10-gauge, four-conductor cable

3¼-in.- to 3½-in.-deep box

Ground

Hot

Neutral

30-amp, four-pole, three-wire grounding receptacle

Instead of mounting a separate washer and dryer receptacle, opt for the ones that are integrated with the hot/cold/drain plumbing assembly. This saves time and money and places the outlets at a convenient location.

yourselfer replaces a blown element, he or she may accidentally replace the 3,500-watt element with a standard 4,500-watt element. The large element will overstress the 12-gauge cable with excessive current and could start a fire. The same thing will happen if the water heater is replaced with a larger 4,500-watt unit. Therefore, be sure to write on the water heater—in large letters—that the 4,500-watt element has been replaced with a 3,500-watt one. Personally, I don't think it's worth the risk. For that reason, I recommend using 10-gauge wire.

Heat pump

Heat pumps are normally wired by their installers, but sometimes an electrician needs to run the feed wires to the unit. If so, the installer tells the electrician the size of the wire needed for the system. There are usually two locations that need to be fed—one outside for the air handler and one inside for the main unit. The wire I use most often for the outside air handler is 10-2 NMB w/g. The size of the cable for the inside unit depends mostly on the size of the electric backup heating.

Disconnects for new installations are required at both the outside and the inside unit if one is not already in the unit. If the inside unit is in an isolated area, it needs switched lighting and a receptacle outlet nearby for maintenance and troubleshooting.

Electric dryer

Dryers are the most miswired of all the appliances because they are both 120 and 240 volts. And it's all because of the neutral and ground. Until 1996, the NEC allowed the neutral and ground to be the same conductor. This put the frame in the conduction path. Although this configuration still had a good safety record, it is now a code violation. Therefore, a dryer should always have a three-conductor cable—three insulated hot wires and a ground— such as 10-3 w/g NMB.

The design load must be at least 5,000 watts or the nameplate rating, whichever is higher. For a standard 240-volt electric dryer, that's a minimum of 21 amps of current, so the rough-in wiring must be 10 gauge and the circuit must have a 30-amp breaker. The wiring terminates in a four-slot (ground, neutral, hot, hot) receptacle.

Garage-Door Openers

Although most garage outlets must be GFCI protected, the outlet for the garage-door opener does not because it's normally located overhead, next to the opener itself. However, I recommend using a GFCI in this instance. People have been known to use the outlet for general purposes, even if they have to get on a ladder to reach it. In addition, a GFCI may help keep lightning off the opener's circuit and protect its electronics.

The receptacle and opener must be installed directly behind the end of the proposed door location in the upright and raised position. That means you can't install the rough-in wiring for the opener until you know the height of the garage door. For example, my garage has an 8-ft. ceiling and a 7-ft. garage door, so I mounted the opener 10 ft. back from the wall where the garage door was to be installed and powered it from a GFCI receptacle mounted next to the opener.

Wiring a "Two-Wire" Submersible Pump

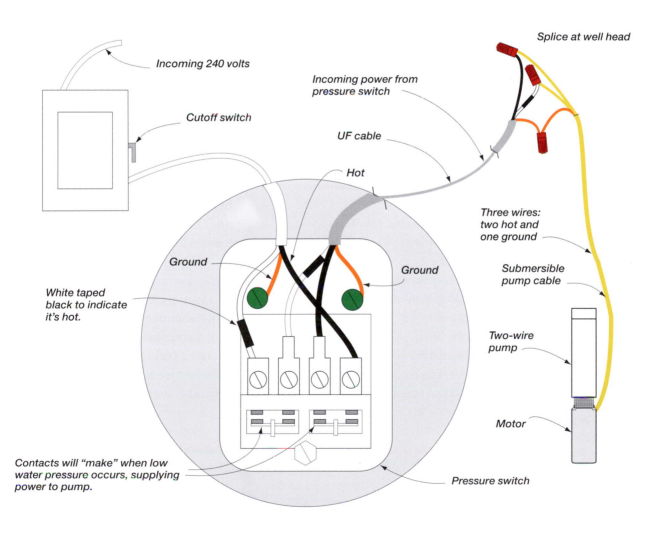

Incoming 240 volts

Splice at well head

Incoming power from pressure switch

Cutoff switch

UF cable

Hot

Three wires: two hot and one ground

Ground

Ground

Submersible pump cable

White taped black to indicate it's hot.

Two-wire pump

Motor

Contacts will "make" when low water pressure occurs, supplying power to pump.

Pressure switch

When installing the water line and UF cable from the house to a well, place it in black corrugated pipe (use pipe with slits or holes so water can drain out). The flexible pipe comes in both 3-in. and 4-in. diameters. The pipe protects the line and cable from rocks and allows them to be easily replaced without having to completely retrench the yard.

In certain instances, a pump's wire chart may indicate that you can use 14-gauge cable. Do not be tempted to do so— 12 gauge should be the smallest cable you use. The reason is you may have to install a higher-horsepower pump at a later date (for example, your neighbor drills a well and the water table lowers). If you install 14-gauge cable, you may need to replace all the cable, which includes a new trench to the well.

Submersible Pumps

Submersible pumps come in many horse-power ratings and in two wiring configurations: a two-wire system that needs only 240 volts and a three-wire system that needs a control box to provide a start and run wiring to the pump. Whether you use a two-wire or three-wire system, it is code to ground the pump and the well casing, if it is metal.

The wiring to a well must be approved for direct burial. It cannot be the standard yellow submersible-pump wire or even the twisted wire so commonly used within a well. In other words, use common UF cable from the pressure switch or control box to the well. Conversely, the wiring within the well should be listed as pump wire, not UF cable. The gauge for both cables depends on the depth of the well, the distance to the well, and the horse-power of the pump. A 140-ft.-deep well located 100 ft. from the house and powered by a standard ½-hp pump can get by with 12-gauge wire. A wire chart in the pump's instruction manual lists the gauges required.

It does not matter whether you decide to use a two-wire or a three-wire pump—they both work just fine. I recommend two-wire pumps because they're easier and cheaper to install (no control box installation and one less conductor in the cable). For a two-wire pump, the branch-circuit wiring terminates first at the pump's cutoff switch and then jumps to the pressure switch. From there, run UF cable to the well casing, where the splice is made under the well cap to the pump cable. This splice can be either insulated crimp (heat shrink) or wire connector. If using wire connecter, point the skirts downward so that condensation does not pool in the connector.

Do not match the breaker size to the cable gauge run to the submersible pump. Many times a large-gauge wire is used to lower the resistance in the circuit on long runs to a well. For instance, for a ⅓-hp pump (a 4-amp load), do not install a 30-amp breaker just because you ran 10-gauge wire to reach the pump; instead, use a 15-amp breaker. The smaller the breaker, the faster it will kick if something happens.

Whirlpool Tubs and Spas

Whirlpool tubs need to have their motors powered through GFCI-protected receptacles or breakers. Some whirlpool tubs come with motors that are connected with a cord and plug. Bring the cable, normally NMB 12-2 w/g for whirlpools and much larger for spas, to the motor and then install a GFCI circuit breaker at the main panel. The receptacle and the motor are maintainable items and need to remain accessible. Remember this, because it may affect the way the tub is oriented in the room. Small whirlpools come with a built-in skirt and access panel. Larger units must have a skirt and an access panel added on-site.

Spas also need GFCI protection. Sometimes the protection is included with the spa; other times, it is the responsibility of the installer. If it is the latter, you may have a problem. High-amperage GFCIs are expensive and can be hard to find.

Always install a dedicated circuit for whirlpools and spas. If either one is hard-wired and a disconnect is not on the unit, you'll need to install it. For a spa, you'll also need to install an emergency cutoff switch and label it prominently.

CHAPTER 13

Lightning and Surge Protection

L O V E   L I G H T N I N G . I've always been fascinated by the way the bolts crisscross the sky. Something so powerful and uncontrollable can play wonders with a child's imagination. And I've never grown up.

Although I've never lost my awe of lighting, I've had to deal with some of its—pardon the pun—negative effects. Protecting yourself and your home from the effects of lightning is not a big deal. If you understand how the surges get into the house, it's an easy jump to adding equipment to keep them out. The trick here is choosing the right equipment (I've installed a lot that didn't work). You can learn from my mistakes. In this chapter, I'll tell you what has been field-tested and actually does what it claims.

Induced Voltage

We all know that lightning is dangerous up close, but I had no idea that its magnetic lines of force were powerful even at a long distance. I was first introduced to lightning's far-reaching effects in the

Induced Voltage

Cloud-to-Cloud or Direct Hit through Utility Lines

Directly into House

mountains of Virginia. I was checking newly buried telephone cable to make sure the plastic jacket hadn't been ripped during installation. Thus, the cable ends were just sticking straight up out of the ground. Thunder rolled far off in the distance but I thought it was too far away to affect me—I thought wrong. As I touched the ungrounded cable shield, it zapped me hard. Lesson number one: The magnetic lines of force from distant lighting can build a charge onto the cable's metal shield (about a mile of it in this one section) and it can discharge through a person to ground.

Since it had already discharged once, I thought it would be safe to touch it again—and it zapped me again. Lesson number two: Drain the shield with a jumper and give it a longer time to discharge the static voltage buildup.

I then ran a jumper from the shield to earth, waited a couple of minutes for the charge on the shield to completely drain away, and then removed the jumper. As I touched the shield a third time, it bit me again . . . hard. Third lesson: Lightning can instantly rebuild a charge on a long, continuous conductor from very far away.

Lightning three, Rex zero. I went home.

The invisible lines of force that I encountered on that cable are the same ones that can wreak havoc on the conductors and appliances in your home. This is why it is imperative to install surge protection on the current-carrying conductors and ground all noncurrent-carrying conductors. A noncurrent-carrying conductor is any metal that isn't used as part of the wiring system but can conduct electricity, such as an appliance frame, steel beams, furnace duct work, metal water or drain line, and metal flues. Connect a #4 copper wire from what needs to be grounded to the

grounding electrode (normally a ground rod), to the grounding electrode conductor (the big copper wire going from the service panel to the ground rod), or to the service panel grounding bus.

Direct Strikes

No matter how "biting" the long distance effects of lightning may feel, it cannot compare to close strikes. Massive amounts of induced voltage from lightning can build up on every wire, metal appliance frame, and metal water pipe in your house just as it did on that telephone cable shield. That's one of many reasons, as we will find out later, that all such items in your house need to be connected to a low-resistance grounding system. Picture the grounding system as a drain—one that takes away current rather than water. If the drain isn't working properly, the electricity can arc over and cause electrical damage as it finds another path.

In addition to lightning-induced voltage saturating your house, which can create voltage spikes everywhere, surges can enter via the utility expressways: the power, coaxial cable, and telephone lines. Massive spikes can run into your house, look around, and find all kinds of damage to do . . . and then zap! Fortunately, it is possible to limit the zaps that come in on the utility lines by clipping off their highest peaks and sending them to earth.

The neutral, being grounded at various poles and at the utility transformer, is thought by many to have no pulse coming into the house. In my opinion, this is not true. The theory that the neutral pulse voltage is zero stems from the assumption that the ground resistance provided by the utility to drain away all neutral surges is 0 ohms. But those who have ever measured ground resistance know this is not true (5 ohms is considered very low, less than 50 is considered adequate by some, and 100 ohms or more is considered high). In truth, there is no such thing as a perfect ground. We drive one or more ground rods, tie into whatever else is considered grounded, and live with what we get.

Utility and In-House Surge Creators

Power utilities have to work on their lines. They are always adding this, fixing that, installing new poles, replacing broken ones, tapping into lines for new service, and so on. Each time they come on-line (and sometimes off-line as well) a surge is placed on the utility conductors. For example, a capacitor bank (installed to reduce voltage fluctuations) produces a significant pulse and initially charges as it is brought up on-line. Trees falling on the lines also create voltage and current surges that can be as damaging as lightning.

The equipment we try to protect also produces surges that can hurt other equipment. For example, a washing machine solenoid snapping in and out sends pulses on the in-house electrical wiring. These pulses may not be as massive as a lightning surge, but they are a lot more numerous. Sometimes the problem is so simple that we overlook it (such as a common snap switch). If you can help it, never put a snap switch on the same circuit as an electronic device.

Here is a simple test: Plug a radio in to the outlet and turn the room lights on and off. The crack you hear coming from the radio is the pulse created by the switch. If you have a computer on that circuit, you are sending a pulse to it as well.

In the past, the only way to ensure that appliances were safe from damaging surges was to unplug them. Back then, most people had only two major appliances: a stove and a washing machine. Not too long after came electric dryers and television sets. Today, unplugging appliances is no longer practical. Our appliances are integral with the house, but they are also very fragile. Operating at only a few volts, they can't survive a battle with lighting or any other surge source. This cause and effect brings on the newest step in home evolution—surge protection.

Adding Surge Protection

A house needs surge protection in two locations: at the point-of-entry (the main panel or subpanel) and at the point-of-use (where the equipment is used). Point-of-entry devices will have a longer life if you use a heavy-duty protector with a high clamping level that affects only large pulses. If a super-sensitive system is installed at the point-of-entry (say, one that cuts and filters every surge down to just above a nominal 120 volts), the protective device will degrade and die an early death from the excessive number of times it must divert surges. Let the point-of-use devices deal with the smaller pulses.

Point-of-entry protection

This is the first line of defense against surges. Without a point-of-entry protection system, one voltage spike can affect the entire house. This happens through the service panel. The service panel is the central hub, and every wire, receptacle, switch, and load is connected to this

Surge Protection

SEU

Main service panel

Stage 1 protection: Point-of-entry protection, surge-breaker arrester, or panel-mount arrester

Stage 2 protection: Plug-in-type point-of-use protection

panel. Although this may be an efficient system for wire distribution, it also allows any incoming surge to branch out and run on every circuit throughout the house.

A surge can come into the panel from the utility (as in a lightning strike) or from in-house equipment, such as a washing machine solenoid or switching appliance. But no matter where it comes from, once a surge hits the main panel it will go through the house like a virus, infecting all the circuits. The only thing that can stop it is an effective point-of-entry protection device. There are two basic types of point-of-entry protection: plug-in and hardwired protection.

PLUG-IN SURGE PROTECTION Plug-in surge protection is the easiest type of protection to install—and, perhaps, the best.

No Entry Protection

Without point-of-entry protection, every surge that comes in on the utility goes through the service panel, out to all the branch circuits, and to the loads.

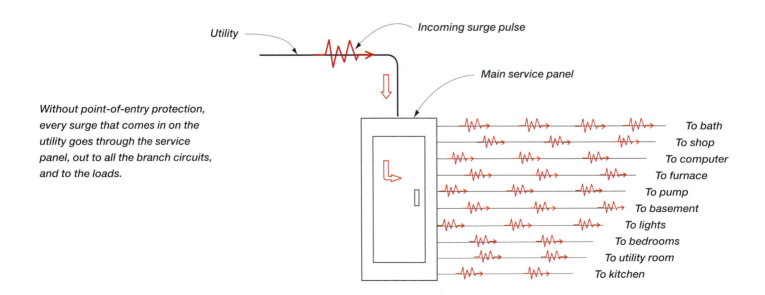

Utility

Incoming surge pulse

Main service panel

To bath
To shop
To computer
To furnace
To pump
To basement
To lights
To bedrooms
To utility room
To kitchen

Noise Pulses

Surges and noise pulses can flow from the load to the service panel. Without point-of-entry protection, pulses and electrical noise generated in-house, such as those from a vacuum cleaner, washing machine solenoid, or shop motor, flow from the branch circuit on which they originate to the main service panel. From there, they flow onto every branch circuit in the house.

Utility — Main service panel — Noise pulses — Utility room — Noise pulses — Shop — Branch circuits — Noise pulses

ABOVE CODE The surge breaker, officially called a Circuit Breaker and Secondary Surge Arrester, made by Siemens, is what I use in my Above Code system. It consists of two single-pole breakers (to protect each phase) and installs just like a common circuit breaker. A red light indicates whether the protection system is working. The nice thing about this system is that it protects the entire panel.

A surge breaker has a tremendous advantage over a hardwired arrester because it snaps right onto the tabs of the service panel and has no long leads that may develop excessive voltages caused by massive surge currents (see Hardwired surge protection below) taken off the line. It has just one wire connecting the breaker to the ground bus. This wire must be as short as possible.

If there is more than one breaker box (say, two 200-amp boxes), you need a surge breaker in each box. If there is a subpanel downstream from the main

This two single-pole breaker module, made by Siemens, mounts in the main service panel like a standard breaker, but it also works as a surge arrester to protect the entire house. The lights indicate that the surge-protection system is working.

panel (in a different room), you should put one in there as well.

HARDWIRED SURGE PROTECTION
Hardwired protection is exactly what it sounds like—the wires are stripped and connected into some type of device. These are further broken down into two subcategories: those that are encased

Although this Primax mounts outside the panel, it is encased in a metal box. The hardwired surge protector mounts on the side of the breaker panel and clips off surges coming in from the utility line. It receives power from a 30-amp double-pole breaker. Do not hide these within the walls.

within metal boxes (either one of its own or the service panel itself) and those that are not. You want the former, not the latter. Any point-of-entry surge protection device that takes massive surges through its body has the potential to explode. (I have seen the damage caused by the resulting shrapnel.) For that reason, hardwired, point-of-entry protection must be mounted either within a metal service panel or within its own metal box adjoining the panel.

When mounted on the outside frame of the panel, make sure that the surge protector is not hidden within a wall, which makes it inaccessible for inspection. (I don't like the idea of having the device explode inside a wall where it could start a fire.) An example of a well-designed, hardwired system is the Primax™ unit made by Panamax. It is encased in a steel box, mounts immediately next to the service panel, and is ideal for surface-mounted panels. For flush-mounted panels, mount the unit on the wall immediately adjacent to the panel and run the wiring into the panel via a nipple on the Primax's side. This particular unit also has an audible alarm to indicate loss of protection.

Point-of-use protection

Point-of-use protection is the second stage of surge protection. Once the massive surges have been taken off the line at the service panel, the more sensitive components of point-of-use protection can come into play.

Point-of-use protection normally comes on a receptacle strip with a long, plug-in cord. The specific type of point-of-use protection you need depends on the equipment you want to protect (AC, telephone, cable, satellite, and so on). You need point-of-use protection for each appliance or group of equipment you want to protect. For example, I have two in my office behind the computer to protect it and all the equipment around it. Then I have another at the portable phone in the kitchen, one for each of the two TV sets, one for the copying machine, one for the stereo system, and so on.

Not installing point-of-use protection shortens the life of all your expensive electronic equipment and entertainment playthings. It is folly to go out and buy a $2,500 top-of-the-line

Keep the Wires Short

The length of the wires on an arrester may dictate whether the unit will work or not. Massive surge currents flowing through these wires develop a voltage proportional to their length. A wire with the resistance of only 1 ohm is still subject to Ohm's Law: Voltage = Current × Resistance. For example, 10,000 amps flowing through a 1-ohm-resistance wire develop a whopping 10,000 volts...and that's enough to blow anything apart.

The rule to remember here is the shorter the wire, the smaller the voltage. Always mount a hardwired device as close as you can to the main service panel. Once the leads enter the panel, keep the leads as short as possible, with no extra slack wound in the panel.

digital-ready TV set and not spend a few more dollars to protect it against surges that will either burn it up or hasten its demise.

AC protection comes in various-size outlet strips, including two, four, six, and eight plug-ins. My experience is that you need at least six; eight is better. Make sure the strip is designed to allow room for transformers to be mounted side by side or back to back.

PICKING THE RIGHT PROTECTION

The biggest challenge in selecting a point-of-use protection device is choosing from such a large assortment of products. Unfortunately, some of the devices on the marketplace are just junk and others are a fire hazard. Here is a list of prerequisites to help you separate the wheat from the chaff:

- UL approval.
- Automatic power disconnect. If the unit fails, which can happen when a massive surge or electrical fault occurs on the line, the unit will disconnect all power to the load.
- A self-diagnostic system. Look for units that indicate whether the receptacle is properly grounded and the protection system is operating.
- An abundance of receptacle slots for all the AC, phone, and coaxial cable lines you want to protect. Make sure it can house several transformers at once.
- A good warranty and guarantee. In addition to insuring itself, look for a unit that insures the objects it's protecting. Some come with a $100,000 insurance policy.

PROBLEMS WITH POINT-OF-USE PROTECTION
Nothing is without problems, and Murphy's Law seems to work overtime with anything electrical. Here

are the most common problems I see in installing point-of-use protection:

- Poor or nonexistent grounding. No low-resistance grounding means no protection.
- Wrong surge device used. Having an unprotected appliance, such as a telephone line or a coaxial cable to the TV or VCR, endangers the entire system.
- Ground skew. This situation is also caused by using two or more different circuits for interconnected equipment. In the drawing on p. 234, there are two ground references. This could cause a difference in ground potential and even a ground loop (current flowing on the ground wires between circuits).

Missing Protection

The AC is protected but the coaxial cable is not. You can have the same problem at a computer by not protecting the phone line.

Coaxial cable is not routed through the protector.

VCR

AC

TV

AC

Signal

Route coaxial cable to the protector and then to the VCR.

AC

Point-of-use protector

Ground Skew

Here, interconnected appliances—the computer and the printer—are plugged into two different circuits. The printer is not protected and it has a different ground reference from the computer. This is known as ground skew. As a result, the equipment may not work correctly at times.

Phone data

Data

Data

CRT

Printer

PC

AC

Circuit #1

Circuit #2

Wrong! Must be on circuit #1.

Point-of-use protector

AC

■ Used outside, around water, or in high heat. Keep any point-of-use device (or anything electrical, for that matter) away from water and high humidity. Also, don't put the unit next to a wood burning stove, which you might do to protect the blower.

■ Connected to a two-prong receptacle through a cheater plug or by cutting off the ground lug. Again, no ground means no protection.

■ Connected to an outlet that uses cheap push-in connections. This type of receptacle overheats easily and can produce intermittent connections. Rewire the receptacle by putting the wires under the screws, or even better, replace the receptacle with a good-quality one.

Phone and Coaxial Cable Protection

Surges coming into the house via the phone and cable lines are quite common and problematic. They destroy computers, TVs, VCRs, modems, answering systems, telephones, and just about anything to which these lines are connected. Although both aerial and buried cables are shielded to protect the wires inside them from lightning-induced surges, the system obviously is not without problems and can pick up major surges. One problem is that many utility systems do not have enough ground rods at their poles and pedestals.

For the phone system, another problem lies with the network interface device (also called NID, or a network interface unit). The NID is mounted on the house siding and is where the telephone cable comes in and the house telephone wire goes out. It provides catastrophic surge protection and an interface point between the house telephone wiring and the utility cable. Back in the old days, an NID was just a tall metal can, and lightning protection was provided via carbon blocks or fuses. Except for providing protection from massive strikes, they did not work well. But then they didn't have to—our telephones were not as sensitive as they are now.

As the electronic era emerged, better methods of telephone surge control were needed. The carbon blocks were replaced with gas tubes, but the first models were a bit too sensitive and put a lot of noise on the line. Today's NIDs work best for long-duration massive surges (I've seen many a protector box literally blown off the house), but they aren't designed to protect your in-house equipment from quick, small surges. It's up to you to provide your own protection. And, as with AC protection, phone surge protectors start with low-resistance grounding.

Telephone protection

You can protect your own phone and cable system by simply adding point-of-use protection (always needed) and point-of-entry protection (optional; you'll have to use a commercial unit). Most residential and small- or home-business installations don't have thousands of feet of telephone lines running throughout the building and can normally get by with point-of-use protection for each piece of equipment or line. For extremely large homes and businesses, I recommend commercial point-of-entry protection in addition to point-of-use protection.

Coaxial cable protection

Most houses have some type of coaxial cable, either from their own antenna or satellite or from the utility. Residential cable point-of-entry protection doesn't exist unless you use a commercial product. The blocks that the cable connects to on your house are grounding blocks, not surge-protection blocks. Their purpose is to send to ground any noise or surges that their shields pick up.

In some areas, installers have stopped putting in grounding blocks and simply use them as a coupling. There are two reasons for this: First, installers don't want to go to the trouble and expense of running the ground wire from the block to the other side of the house where the grounding system is located. Second, the runs from the dish to the house equipment are sometimes very short and don't have the potential to pick up surges as longer runs do. But this logic is thrown out the window as soon as a lightning strike hits a nearby tree and all your equipment goes up in smoke.

Don't assume that the equipment the coaxial cable connects to after it enters the house will ground the cable's shield and thus take the surges to ground. Most electronic equipment is two-wire AC (no ground). The bottom line is that without point-of-use protection to provide both protection and grounding, the coax shield will remain ungrounded, waiting for a surge to burn up the electronic gear.

Point-of-use protection for both phone and cable is quite easy to obtain. Many multioutlet point-of-use AC line protection systems safeguard these circuits as well. The most unique and versatile concept in protection design that I've seen is called the MAX 6 (made by Panamax). This plug-in protector has a base module providing AC protection and removable modular ends for different types of wiring protection, such as coaxial, telephone, and data lines.

Protecting Pumps

In high-lightning areas, submersible pumps can easily burn up even if they are under water. Protecting pumps from lightning is a little different from protecting appliances within the house. In addition to surges coming down the utility lines, pumps are bombarded with heavy surges from a variety of directions.

After years of experimenting with brands and designs, I finally came up with a system that worked. (My apologies to those who received the systems that didn't.) The first part of my system is a Siemens surge breaker or a hardwired Primax unit in the main service panel to stop massive surges from coming in on the utility line. In high-lightning areas, not having one of these means certain death for a submersible pump.

How Lightning Finds a Pump

Lightning can destroy a submersible pump by entering in the opposite direction. For instance, lightning hits a tree and finds its way through the ground to the pump—sometimes through the earth, sometimes by running a fence. By the time the surge gets to the protector in the panel, the pump is toast.

Lightning runs fence and jumps into the well.

Lightning goes through roots to the well.

This burned hulk is what's left after a surge came in backward from a well. The surge-protection system within the pump protected the motor but not the control box (the surge also burned out the pressure switch). I've seen the lids of these boxes explode across a room.

Pumps can get major hits from other sources as well, especially when the pump system is a long distance from the service panel. Typical problems include induced surges into the pump wire from overhead lightning, strikes hitting a metal vent line and jumping to the pump wire, strikes running along metal water lines, lightning running along fence lines and jumping to the pump wire, and lightning hitting a tree and running into the ground to the well. Thus, a surge comes into the house backward and blows apart the first thing it sees—namely, the pump control box and pressure switch. (Both of these units are located at the water pressure tank.)

The control box starts the motor in three-wire pumps (two-wire pumps have this circuitry in the motor) and the pressure switch turns the system on and off as water is needed. I've actually seen control boxes that have blown across a room and punched holes in the drywall. This is another reason why I started wiring in the Primax unit at the cutoff switch, control box, or pressure switch.

For wells nowhere near the house, pumps need protection at the wellhead. Here I use a hardwired arrester cut right into the wires and placed under the well cap. I use a Tytewadd suppressor unit because these wells are normally in some type of building housing the electrical gear.

My last piece of magic is an Above Code grounding system to make sure surges always have a low-resistance ground. Be sure to include the metal well casing in the grounding system. If the well is close to the house, run a #4 copper wire from the last ground rod to the casing.

A submersible pump deep in a well has its own protection in the motor, but it still can be damaged. When the surge comes in to the pump's motor, the pulse races to the end of the winding and doubles in value each time it bounces back to the opposite end of the winding—10,000 volts, 20,000 volts, 40,000 volts—until the windings fail.

Many pump motors don't fail because of just one thunderstorm. The damage is cumulative, with the windings becoming more susceptible to catastrophic damage with each surge until the unit just dies one day. Without a breaker surge arrester mounted in the cutoff switch to limit the surges before they reach the pump motor, lightning surges stream down into the motor time and time again and overstress the motor windings. The motor may continue to work, but the day will come when it takes one surge too many—or just gives up the ghost during Thanksgiving or Christmas.

Well Casings

If you are having a well drilled, the installer will give you a choice of casings: galvanized steel, nongalvanized steel, and plastic (PVC). Always choose galvanized steel. Nongalvanized steel can rust and contaminate the well. PVC is less expensive and doesn't rust, but it has two big disadvantages. First, you lose the ability to include the casing in the grounding system, which makes the absolute best ground possible. Second, PVC doesn't seal well in bedrock. If you use steel casing, the driller can pound it hard into the rock for a good, watertight seal. You cannot do that with plastic, because it will shear off and block the casing hole.

Protecting a Pump

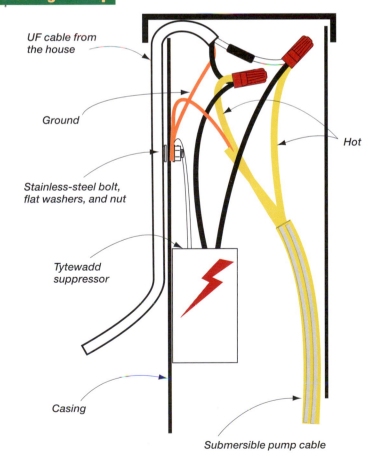

UF cable from the house

Ground

Stainless-steel bolt, flat washers, and nut

Tytewadd suppressor

Casing

Hot

Submersible pump cable

Lightning Protection for Pumps

One of my customers went through three submersible pumps in three days—all in the same storm.

As I was replacing the original pump (between rain showers), I told the customer of my multistage system of lightning protection for pumps. He thought it sounded great, but he didn't want to buy it right then and there. The next day he called me and said he had lost the new pump I had just installed—again to lightning.

As I was replacing the pump for the second time, I again suggested my special system. This time he agreed, but said that he would install it himself in order to save on labor. I sold him the material, told him what to do, and went on my way.

The very next day, just like clockwork, he called again (by now we were becoming close friends). He had lost the second pump to lightning. I asked him if he had installed my special system because I had never lost anything with it installed properly. He said he hadn't gotten around to it yet, but he would now. After I installed the third pump and he installed my multistage pump protection system, I never heard from him again (so much for good friends). Bottom line: Put the system in before you lose your pump, not after.

Pump Protection System

A pump lightning-protection system is done in layers just like all other protection systems. The first layer is at the service panel and the second is at the pump controls. The third (optional) layer is at the wellhead.

Use a Siemens surge breaker or a Primax hardwired system in the main service panel.

Wire in a Primax unit at the cutoff switch adjacent to the pressure switch or pump control box.

Install an Above Code grounding system by including a ground to the metal well casing.

Optional: If a well casing is farther away than 100 ft. and there is a danger of lightning directly hitting the well from a tall, deep-rooted tree or a fence line, install a heavy-duty, hardwired suppresser (such as a Tytewadd) under the wellhead.

Layer #1: Required protection at main panel

Power cutoff box (if any)

Layer #2: Primax

Pressure switch

Protection required in one of these three units

Main service panel

Above Code grounding system

Control box (if any)

To pump (optional protection at wellhead)

CHAPTER 14

Standby Generators

EVEN THOUGH Y2K is long gone, the popularity of standby generators is here to stay. Country folk are no strangers to power outages. Miles of power lines run through forests, where trees and limbs constantly fall on the lines. Many of us deal with ice storms that can disrupt power for weeks and winds that can topple power poles like a collapsing row of dominos. And who knows what type of service we will have when the power companies really start competing against each other?

Just as there are no shortages of people wanting standby power, there are no shortages of those who do not know which generator to buy or how to install one properly. Buying the wrong-size generator can be a waste of money (one of my customers bought three before he acquired the right one.) After selection, the next hurdle is installation. An improper installation can be dangerous to both the utility worker and the homeowner and can even cause the generator to go up in flames.

History of Incorrect Connections

The first affordable household generators were small. To use them, people ran extension cords from the generator to the appliances that needed power. With the advent of newer, more powerful generators, people began experimenting with ways to plug in the generator

This on/off switch is not legal. Two-position switches can put voltage on the utility line that can disrupt the utility power and electrocute power-company workers.

Power Tester

To know when the utility power comes back on, place a lightbulb and fixture (such as a 20-watt refrigerator bulb) next to the main service-panel box on a circuit that is always powered by the utility. When you see the light on, you'll know power has been restored and your generator can be turned off.

directly to the house wiring via a household outlet. This method, called back-feeding, is both illegal and dangerous; unfortunately, it became an all-too-common occurrence. This direct-to-receptacle method allows power to run backwards out of the house and through the utility transformer, and it can electrocute the workers repairing the line. When the utility brings the

power back on-line, the generator will go up in flames.

Backfeeding evolved into using switches to turn the power on and off. To keep the generated power in the house, the homeowner would pull the main fuses or turn off the main breaker. A large on/off switch then supplied power to the house. This system is almost as bad as the extension-cord-plugged-into-the-outlet method. If someone accidentally turns on the switch when the utility voltage is coming into the house, sparks will fly, brownouts will occur down the utility line, and the generator will go up in smoke. We cannot have the lives of utility workers dependent on the whim of every homeowner who might accidentally throw the switch. Thus came the need for a special switch that will automatically disconnect the generator circuit from the house the moment the main utility switch is turned on—a three-position switch with the OFF position in the middle.

As previously mentioned, you cannot use any switch, even if it is already installed and looks official, that does not have a three-position "ON-OFF-ON" handle (it could also say "LINE-OFF-GEN"). The middle OFF position is crucial. It is the part that prevents you from cross-connecting the generator with the utility.

In the past, the standard legal way to connect a generator to the house was to install a transfer switch between the meter base and the service panel. This system no longer meets code. You now have two options. You can wire a separate panel adjacent to the service panel to contain all the circuits powered by the generator (which is a royal expense and pain) and feed it into the service panel. Alternately, you can install a quick-and-easy transfer switch akin to the Gen/Tran® system.

If wired into the house, a generator must be controlled through a three-position switch with the middle position being OFF. This can be done via individual switches, such as with this Gen/Tran unit (above), or as one giant switch (below).

Picking a Generator

Wattage, wattage, wattage. That's all most people think about when they pick out a generator. Wattage is important. Without the proper wattage you can't power the house or an appliance. However, wattage is not all-important. There are many other variables that must be considered before you spend your hard-earned cash.

Wattage and circuits

To figure out the wattage you need, think about all the appliances you want to power. And while you're doing it, remember that each appliance may be an individual circuit unto itself and you are limited to the number of circuits you can power. Thus, your two limitations: wattage and power. Realize that you will not be able to power everything. Make an emergency list of appliances that simply must have power. The freezer and refrigerator are the two most common. After that come a few lights and water (if you have a pump). Hot water (for electric water heaters) is hard to obtain unless you buy a very-high-wattage generator.

Rather than adding up large lists of wattages, you can buy a basic 4,500-watt generator. This size generator is large enough for almost any short-duration emergency situation. If you want more wattage, by all means go for it, but I don't recommend anything smaller than a 4,500-watt unit.

Some appliances need a "kick" to start and thus require a little extra wattage. Most compressors and motors need two to four times their run wattage in order to start. For example, a ¾-hp submersible pump needs a 3,500-watt generator to start pumping water into the house. Knowing this, many generator manufacturers have surge current capabilities built into their generators. Some unscrupulous salesmen will try to hype

This generator proudly advertises 6,300 watts. But to find the real wattage, read the nameplate on the generator. It's 5,900 watts—a big difference.

the surge wattage, which is usually 200 to 500 watts greater than the run wattage, as the run wattage. Don't be fooled. Surge wattage is available from the generator for only a few seconds.

For every appliance you "had to have," you also added one circuit for each 120-volt appliance or load and two circuits for each 240-volt appliance or load. These circuits add up fast. In most cases, go with 6 or 10 circuits for the Gen/Tran. An average house with ordinary loads can get by with six emergency circuits, which is all that a typical generator (for example, the 4,500-watt unit mentioned earlier) can power. Typical emergency loads and the number of circuits for a house may include:

- Water pump; two circuits
- Water heater, if any; two circuits
- Lights and receptacles; one area per circuit (combine light-duty load circuits, if needed)
- Refrigerator; one circuit

Two House Circuits for the Price of One

It's possible to have two individual house circuits powered by a single Gen/Tran circuit by rearranging and combining circuits in the service panel. For example, if you have two bedrooms with no heavy-wattage appliances, you can combine them as one circuit in the service panel for the Gen/Tran. Simply splice the two together with a wire connector, and then add a jumper to the splice for the Gen/Tran. You'll now power two circuits for the price of one. Remember to spread the compressor load appliances on different GenTran circuits—don't wire them together or they will kick the breaker when both start at the same time.

One turn of the key and your generator is up and running. You'll either get an electric start or wish you had. The hand pull is always there as a backup.

- Freezer; one circuit
- Computer or living room; one circuit

For more circuits, you can jump up to a 7,500-watt generator and get a 10-circuit panel. You can use a 10-circuit panel with a smaller 4,500-watt generator, but you'll have less wattage to spread across the number of circuits. If you are building a new home, design the house wiring so that you can install a maximum number of circuits into a 10-circuit Gen/Tran.

Tank size

You want a large fuel tank—one that provides the generator with enough fuel so you don't have to keep shutting the system down every hour or two to fill it with gas. But just getting a large tank isn't enough. The actual time between fills depends on a variety of factors, including the motor's horsepower and efficiency. Check the generator's spec sheet for the "run time," which takes all of these factors into consideration.

You will also want a way to tell how much gas is left in the tank without having to stop the engine, open the lid, and insert a stick to see if it comes up

wet. Fuel gauges for generators are commonly built right into the lid. If your generator doesn't have one, get one. They are usually sold adjacent to the generator display. (Or even better, have the salesperson throw it in with the package.)

Easy-starting engine

If the generator won't start when you need it, you may as well not have it. Most generators are ignored for months before they are used, and many have hard-starting engines similar to those on some lawnmowers. In my experience, the easiest engine to start is a Honda® engine—it starts with no trouble even when ignored for a year or more. Honda engines cost more, so they're generally used in only top-of-the-line generators.

I always recommend electric-start engines. If you don't get one, you'll wish

WORK SAFE
WORK SMART
THINKING AHEAD

Never fill the tank with the generator running—the gas can ignite if some sloshes over the side or if the fumes swirl over to the exhaust.

you had; hand-pulling gets old very quickly. The problem with electric starts is keeping the batteries charged—they seem to drain quickly during nonuse months and die easily. For that reason, I recommend a large starting battery with about twice the cranking size recommended by the manufacturer. It's also a good idea to start the generator once or twice a year, just to be sure the battery is properly charged.

Dolly or cart

Generators are heavy and noisy. If your generator isn't on wheels, you should consider making or buying a dolly so that you can move it around easily. Because of the dangers from exhaust gases, the generator must run outside, away from open doors, windows, and garage doors, which provide entryways for poisonous exhaust.

Receptacles

Generators come with a great number and type of receptacles, including 120 volt, 15 and 20 amp; 240 volt, 15 and 20 amp; 120/240 volt, 20, 30, and 40 amp; and twist-lock receptacles. Don't let the wide variety worry you. If

The four-slot receptacle is designed to supply power to an entire house. The slots are designed for a twist-lock plug and four-conductor cable. This circuit provides both 120 volts and 240 volts.

Exhaust Fumes Can Kill

A few years back, when a heavy snow took down power lines to his house, a Vermont timber framer started a standby generator that he stored in his attached garage. Even with the garage door open a few feet for ventilation, a draft carried exhaust gases into the house. Wakened to the smell of fumes, the man went to investigate and was overcome by carbon monoxide and died. He left behind a wife and two daughters.

To keep the generator (and its fumes) a safe distance from your house, use a long extension cord to reach from the switch panel to the generator or hardwire a cable from inside the switch panel to the distant generator location. If the cord approaches 100 ft., consider the latter.

you are only going to use the generator to run an appliance or two off an extension cord, then all you need is a 120-volt, GFCI-protected, 15- or 20-amp receptacle on the generator. Almost all generators have that. But if you are going to run an entire house, you will need a four-prong, twist-lock outlet rated at 120/240 volts—meaning 120 *and* 240 volts (which requires two insulated hot wires, an insulated neutral, and a bare ground) all in one outlet.

Installing the Transfer Switch

Installing a manual Gen/Tran-style transfer switch is the easiest way to hook up a generator. The switch is installed next to the main circuit panel and wired directly to it, connecting the generator to the house wiring. I've installed many Gen/Tran switches (including in my own house) and the average installation time for a six-circuit panel is about an hour. You'll spend more time trying to decide how many circuits you need to power.

Wiring a Surface-Mount Manual Transfer Switch

For easy wiring, mount the transfer switch reasonably close to the service panel. The standard Gen/Tran switch simply screws onto the finished wall adjacent to the service panel; a metal flex cable jumps to the panel for the circuits to feed back and forth. The spot at which you want the pigtail to enter the service panel determines the exact location of the Gen/Tran. Be careful to make only gentle curves with the pigtail or you may split open the armor.

Once the panel is mounted, all you need to do is splice in the wires. Although the process sounds complicated, it requires only a few steps.

1

Turn off the power to the service panel. Mount the transfer switch adjacent to the service panel and bring the pigtail through a panel knockout. Attach the large white wire neutral to the neutral bus and the green wire, if any, to the grounding bus. On a main service panel, they are on the same bus; on a subpanel, they are on different buses.

2

Match each red wire to the same-lettered black wire. Eventually, the red wires connect to a breaker, the black wires to the circuit load. Note the lettered positions on the transfer switch for both the 120-volt and the 240-volt circuits. For example, on my six-circuit transfer switch, C and D are on the 240-volt circuit and the rest are on the 120-volt circuit.

Remove the hot (black) wire from the circuit you want powered by the generator. Splice it onto a black wire from the Gen/Tran.

Attach the red wire with matching letter to the breaker. Follow the same procedure for each circuit you want to wire to the Gen/Tran.

Although the wiring can be hardwired from the Gen/Tran switch to a receptacle close to the generator, most systems use a heavy-duty extension cord to connect the male outlet on the Gen/Tran switch to the female outlet on the generator. You can buy these cords from the manufacturer or make your own, as I did with a 10-3 with ground, outdoor-approved line cord.

When you have finished, switch all the Gen/Tran's switches to "LINE" and power up the main service panel; all the house circuits should then receive utility power. To use standby power, start the generator and let it run for a few minutes, then throw the Gen/Tran switch from "LINE" to "GEN." You will hear the generator lower its rpm as it starts putting out power.

Index